ANCIENT GREEK AGRICULTURE

ANCIENT GREEK AGRICULTURE

An introduction

Signe Isager
and
Jens Erik Skydsgaard

London and New York

First published 1992
by Routledge
11 New Fetter Lane, London EC4P 4EE

Simultaneously published in the USA and Canada
by Routledge
a division of Routledge, Chapman and Hall Inc.
29 West 35th Street, New York, NY 10001

© 1992 Signe Isager and Jens Erik Skydsgaard

Typeset in 10 on 12 point Garamond by
Florencetype Ltd, Weston-super-Mare
Printed in Great Britain by
T J Press (Padstow) Ltd, Padstow, Cornwall

All rights reserved. No part of this book may be reprinted or reproduced or utilized in any form or by any electronic, mechanical, or other means, now known or hereafter invented, including photocopying and recording, or in any information storage or retrieval system, without permission in writing from the publishers.

British Library Cataloguing in Publication Data

Isager, Signe
Ancient Greek agriculture.
 I. Title   II. Skydsgaard, Jens Erik
         630.932

Library of Congress Cataloging in Publication Data

Isager, Signe.
Ancient Greek agriculture: an introduction / Signe Isager and Jens Erik Skydsgaard.
       p. cm.
Includes bibliographical references and index.
1. Agriculture–Greece–History. I. Skydsgaard, Jens Erik.
II. Title.
S429.I73   1992
338.1'0938–dc20

ISBN 0-415-00164-1

CONTENTS

List of Plates and Figures	vii
Preface	ix
Part I The art of agriculture	**1**
INTRODUCTION	3
1 THE GEOGRAPHICAL BACKGROUND	9
2 TILLING AND CROPS	19
3 AGRICULTURAL IMPLEMENTS	44
4 AGRICULTURAL BUILDING	67
5 ANIMAL HUSBANDRY	83
6 AGRARIAN SYSTEMS	108
Part II State and agriculture	**115**
INTRODUCTION	117
7 PRIVATE LAND	120
8 TAXES IN AGRICULTURE	135
9 OTHER LAWS	145
10 LABOUR AND STATE	149
Part III Gods and agriculture	**157**
INTRODUCTION	159
11 THE CALENDAR	160
12 AGRICULTURAL PRODUCTS FOR THE GODS	169
13 LAND BELONGING TO THE GODS	181
14 THE ANIMALS OF THE GODS	191
EPILOGUE	199
Appendix: The sacred olives	203
Bibliography	206
Index of passages cited	224
General index	228

LIST OF PLATES AND FIGURES

Illustrations indicated with (S) are provided by Skydsgaard, the others are supplied by the respective museums. We should like to thank the authorities for permitting reproduction.

Plates

1.1	The Marathonian Plain (S)	16
1.2	Landscape with terraces, Methana (S)	16
2.1	Triptolemus with ear of corn, Attic red-figure hydria, Ny Carlsberg Glyptothek, Copenhagen	23
2.2	Ear of corn, stater from Metapontum, National Museum, Copenhagen	23
2.3	Cuttings of vine, Ithaka (S)	30
2.4	Vineyard just before pruning in January, Methana (S)	30
2.5	Vine, Chios (S)	31
2.6	Wild olive, Athens, Areopagus (S)	34
2.7	Domesticated olive, Methana (S)	34
2.8	Grafting on a wild olive, Methana (S)	37
2.9	The same grafting as in Plate 2.8, six months later (S)	37
2.10	Old olive tree, cut down, with new shoots growing from the trunk, Crete (S)	39
2.11	Picking the olives, Attic black-figure amphora, British Museum	40
3.1	Plough, Boeotian terracotta, Louvre	48
3.2	Modern plough (*ard*), Mykonos (S)	48
3.3	Agricultural labours, Attic black-figure kylix, Louvre	50–1
3.4	Pruning-hooks and reaping sickle, Olynthos	54
3.5	Threshing-floor, Methana (S)	54
3.6	Picking and treading the grapes, Attic black-figure kylix, Cabinet des Medailles, Paris	58
3.7	Treading the grapes, Attic red-figure krater, Museo Civico, Ferrara	58
3.8	Treading the grapes on a stone pressing-bed, Attic red-figure amphora, Museo Civico, Bologna	59

LIST OF PLATES AND FIGURES

3.9	Stone pressing-bed, Olynthos (S)	59
3.10	Oil-mill (*trapetum*), Olynthos (S)	62
3.11	Oil-mill (*trapetum*), Pompeii (S)	62
3.12	Olive-press, Attic black-figure skyphos, Boston (M.H. Hansen)	64
3.13	Olive-press, Attic black-figure skyphos, Boston (M.H. Hansen)	64
3.14	Bags with crushed olives under the press, relief, British Museum (S)	65
5.1	Athenian horsemen, Elgin Marbles, British Museum (S)	86
5.2	Mating donkeys, Attic red-figure oinochoe, Munich	88
5.3	Donkey with packsaddle, cameo, Thorvaldsen Museum, Copenhagen	88
5.4	Herakles leading a bull, Attic red-figure amphora, Boston (after Pfuhl 1923)	90
5.5	Cows in sacrificial procession, Elgin Marbles, British Museum	90
5.6	Goats and herdsman, Attic black-figure kyathos, Louvre (after Pfuhl 1923)	92
5.7	Odysseus escaping under the ram, Attic black-figure lekythos, National Museum, Copenhagen	92
5.8	Going to market with pigs, Attic red-figure pelike, Fitzwilliam Museum, Cambridge	94
5.9	Cocks, Attic Tyrrhenian amphora, National Museum, Copenhagen	95

Figures

1.1	Orographic map of Greece	12
1.2	Hydrographic map of Greece	13
3.1	Diagram of a plough (after Drachmann 1938)	47
3.2	Diagram of olive-mill (*trapetum*) (after Drachmann 1932)	60
4.1	Country house near the Cave of Pan at Vari (Jones/Graham/Sackett 1973)	73
4.2	Small country house near Vari (after Lauter 1980)	75
4.3	Country house no. 26 at the Crimean Chersonesos (Dufková/Pečírka 1970)	77
11.1	The agricultural year and the Attic calendar of festivals	162
12.1	The offering of animals in individual months, Erchia	176

PREFACE

The concept of this book originated some years ago in a Scandinavian symposium on Ancient History. It was obvious that agriculture played an important role in the history of ancient Greece and that there was need of a textbook that could serve as an introduction. Discussions with colleagues have supported the authors in their pursuit of the task and now, with some delay caused by the very disparate work of university teachers, the book is finished.

During our studies we have seen the interest of ancient agriculture grow, resulting in a nearly overwhelming flow of articles and books. Neither of us can say – with Cassius Dio (1.1) – that 'I have read almost everything that has been written by anybody' but we can confirm that 'we have not included all in the treatise'. Selection is difficult and we have decided to concentrate on Greek agriculture of the city-states, that is, from Homer to Aristotle and Theophrastus. The task has been divided. Skydsgaard undertook to write the first, more technical, part whereas Isager has written the latter parts, on the relationship between agriculture on one side and state and gods respectively on the other. We have, nevertheless, collaborated step by step, discussing most of the topics several times. Each of us is therefore responsible for the entire book.

We should like to thank our universities for granting terms free from teaching from time to time. The Carlsberg Foundation and Churchill College, Cambridge, granted Skydsgaard a sabbatical term in Cambridge which was very fruitful, not least because of the hospitality and interest of the colleagues there, including that of the late Moses I. Finley, who kindly encouraged the studies. We also spent a week in Methana as the guests of Lin Foxhall and Hamish Forbes, discussing various aspects of agriculture there, and should like to thank them very much for their hospitality.

The Danish Research Council has kindly given a grant for the translation of the Danish manuscript, and we owe much to Professor Jørgen Læssøe for his translation of the often difficult and technical text. Last, but not least, we should like to extend our thanks to colleagues here and abroad, including

PREFACE

our students, who have all willingly discussed several details of ancient agriculture with us over the last decade.

Signe Isager
University of Odense

Jens Erik Skydsgaard
University of Copenhagen

Part I

THE ART OF AGRICULTURE

INTRODUCTION

'On n'insistera pas ici sur les aspects proprement agricoles (cultures et techniques) de la vie rurale: cela est banal et exposé partout.' Thus Edouard Will in his outstanding history of Greece, *Le Monde grec et L'Orient* (1972). The authors of the present volume have, however, encountered some difficulty in finding the numerous treatises or textbooks on ancient Greek agriculture that Will seems to presuppose. Considering that, today, interest is generally concentrated on agriculture as the most important occupation in a pre-industrial society, we might have expected an increased interest in this occupation, but most authors (such as, for instance, Will) are satisfied with devoting a few, albeit brilliant, pages to the subject. However, agriculture is a complex phenomenon, in history as well as the present day. It requires an intimate knowledge of the natural possibilities and limitations set by climate and soil, and it presupposes the command of a technology that is often very complicated. If, as an industry, agriculture aims at something more than sustaining life within the framework of a family, its production must be viewed in its relation to the needs of the entire community and the economic system. Agriculture is a basic industry, but it does not exist independently, removed from the general norms of the society. It may be argued that from the time when man first began to cultivate the land, agriculture was one of the leading factors in the social structure, primarily because of the status attributed to the land in its various relations to those who worked it. Who owns the land, how is ownership transferred from one person to another, what is the relation between the person who owns the land and the person who works it? A whole series of questions of this nature may be asked, but needless to say not all of them can be answered. Finally, there is the question of cultivation itself: what was cultivated, and how? Here, historians will often find themselves in a difficult situation – cultivated plants may have changed in their essential features over the many centuries separating the present time from ancient Greece. Even in primitive agriculture selection takes place in the reproduction of plants, and it may be exceedingly difficult to envisage the ancient types of grain, let alone undertake quantitative calculations of the yield. Historians may often feel hampered because their botanical insight is limited, and their knowledge of soil science, manuring and many other phenomena, self-evident to the farmer, will frequently be embryonic, to say the least. The work done by the historian is usually tied to ancient sources, and the sources concerning Greek agriculture are extremely scarce, which is paradoxical considering the multitude of sources referring to the importance of agriculture. Greek has no fixed terminology for notions like 'terrace', 'nursery bed', and so on. Should we conclude, then, that these phenomena did not exist? In his classic book, *The Ancient Economy* (1973a), Moses I. Finley has stressed that the absence of a set of terms for economic concepts is due to the lack of economic thinking. Were we to transfer this principle to agriculture, we might be tempted to doubt that it was the main industry of the Greeks. Attempts have been made to fill the gaps in our knowledge in various ways. With H. Michell, *The Economics of Ancient Greece* (1963), one may draw on the Roman sources, first and foremost geographers and agronomists. Here, however, we must take into account that, in essential points, geographical conditions in Italy differ from those of Greece, and that there is a significant chronological difference between Archaic and Classical Greece and Late Republican Italy. It has been argued that technological development in antiquity was extremely slow, if indeed it existed at all. But it was precisely in the Late Republican period of Italy, with its many medium-sized *villae rusticae* under the same owner, that production, in particular of wine and olive-oil intended for the markets, increased. Some of the technical innovations were directly derived from Hellenistic technique, such as the screw press, a prerequisite for which was the Archimedean screw. Even a casual visit to Greece and Italy will reveal to those who are interested that natural conditions for the integration of estates are far more readily at hand in Italy, and if we take a look at economic history, it is immediately apparent that the Roman conquests created an economic upper class that had the will and the means to invest in the traditional industry. Nothing comparable was to be found in Archaic and Classical Greece where the narrow confines of the city-state set a natural limit with regard to the integration of estates and the accumulation of wealth in land. In our opinion, a closer link between the Greek and the Roman sources is a hindrance, rather than a help, towards a more precisely defined understanding of the specific character of Greek agriculture.[1]

A different approach would be to study contemporary Greek agriculture, and to try and make deductions back to antiquity. We find ourselves in the fortunate situation that geographers as well as social anthropologists have taken an interest in the comparatively backward Greek countryside. This is due particularly to political conditions in modern Greece. After the Second World War and the Greek Civil War substantial amounts of capital, primarily American, were invested in order to put Greece back on her feet; it had

---

1 Skydsgaard 1987.

been realized that without a thorough change in the stagnant life in the villages, the demographic distribution of the Greek population would become yet more uneven. A number of commissions under OEEC came up with recommendations supposed to guide governments in their agricultural politics. Thus, in 1951 appeared *Pasture and Fodder Development in Mediterranean Countries*; in general, it deals with the ever-present question concerning the relationship between agriculture and cattle-breeding, emphasizing the need to supply manure to the soil as a prerequisite for an increased agricultural productivity. Subsequent years saw a series of specific studies on Greek villages: *Vasilika, A Village in Modern Greece* by Ernestine Friedl (1962), *Portrait of a Greek Mountain Village* by Juliet du Boulay (1974) and *The Greek Peasant* by Scott G. McNall (1974), with analyses of villages in Boeotia, Euboea and northern Attica. Although the interest of the scholars is focused mainly on social life in a wider sense, much information concerning agriculture is to be found. In 1975 Ernestine Friedl convened a conference in New York with a view to assembling the results of field work in Greece; in 1976 the report was published, entitled *Regional Variation in Modern Greece and Cyprus: Towards a Perspective on the Ethnography of Greece*, edited by M. Dimen and Friedl. Here we find a series of interesting separate analyses which throw light on agricultural practice and farmers' mentality in parts of Greece that are only to a small degree under the influence of the three phenomena that rapidly change the nature of agriculture: the use of fertilizers, artificial irrigation and mechanization. It is also noteworthy that here we find an attempt at a long-range analysis of agriculture. The same trends will be found in the so-called 'New Archaeology'. Here, Michael H. Jameson's project in southern Argolis, which is now nearing its final publication, and the Cambridge/Bradford Boeotian Expedition are of the greatest interest and go to show that a combined effort from scholars representing various fields of learning is of vital importance. The pioneers in Greece were the participants in the Minnesota Messenia Expedition, the results of which were published in 1972 by W.A. McDonald and G.R. Rapp with the subtitle *Reconstructing a Bronze Age Regional Environment*. But the concentration on prehistoric periods may well make the student of antiquity shed a bitter tear. The expedition analysed remains from the Bronze Age, but some may feel that relatively little attention was paid to later remains from the Archaic and Classical periods. A search for the Messenian Helot settlements could perhaps have furnished us with a corresponding result and would have given us an entirely different basis for our understanding of circumstances concerning the production that constituted the basis of Spartan society. Now we have to be content with the impetus that these studies have given to Bronze Age research. But other surveys are on their way. Classical archaeologists have, in fact, overcome the hesitation towards the study of material culture that Sally C. Humphreys, rather caustically, has described in her article 'Archaeology and the social and economic history of Classical

Greece' (1967). The ecological aspect in the interpretation of the past has been emphasized also by Renfrew and Wagstaff in *An Island Polity: The Archaeology of Exploitation in Melos* (1982) and by J.L. Davis, J.F. Cherry and E. Mantzourani in 'An archaeological survey of the Greek island of Keos' (1985). Often, however, it seems that the presentation of archaeological material has receded into the background in favour of interpretation. This lessens the reader's ability to check as he reads on; perhaps, indeed, it throws a veil over the fact that the empirical material is limited, and therefore also of limited value as evidence.[2] It remains to consider another interesting contribution from historical geography: *The Development of Rural Settlement, A Study of the Helos Plain in Southern Greece* by J.M. Wagstaff (1982). By a combination of archaeological and geographical methods we find a history of settlements in a well-defined area on the Peloponnese from prehistoric times until today, with a number of precise observations that require reflection also for a student of ancient history.

These investigations, and others, however, leave us with the fundamental question: to what degree dare we deduce from contemporary conditions to antiquity? Naturally, we are able to make adjustments here and there. The easiest is to disregard crops which we know have arrived in later periods, such as maize, tobacco, citrus fruits and so on. The main work on this subject is still Victor Hehn, *Kulturpflanzen und Haustiere in ihrem Übergang aus Asien nach Griechenland und Italien sowie in das übrige Europa* (1870; reprinted in 1963). It is much more difficult to make deductions from the present landscape to what would have presented itself to us in antiquity. Which phenomena are stable, and which are subject to changes? The geographic determinism that allows one to make unmodified deductions from contemporary pre-industrial agriculture to that of the ancient world holds as many pitfalls for the person who accepts it as it does for one who deliberately rejects it. In its history, post-Classical Greece is marked by a singular lack of continuity of population: so much so that a naïve inference or over-emphasis of constancy inevitably leads us to regard the natural resources as dominant in history; thus the importance of human activity tends to dwindle to a minimum. On the other hand, it is a well-known fact that you cannot grow sugar beets in the Sahara, and those who wish to disregard the later and better-known agrarian history of Greece would do well to abide by this simple rule with all its consequences. One cannot study agrarian history solely on the basis of the surviving literary sources; one must pay attention to the phenomenon known as geographic constancy. The difficulty, of course, is weighing its significance in the individual situation of interpretation.

An account of ancient Greek agriculture requires, therefore, that you steer between Scylla and Charybdis. The interpretation of the sources is, of

---

2 Generally Keller/Rupp 1983.

course, fundamental. The literary sources are readily reviewed. Hesiod's *Works and Days*, traditionally dated to *c.* 700 BC, is important. It consists of a series of very personal reflections on a peasant's life, addressed to Perses, a brother of the poet. The so-called 'agrarian calendar' (ll. 382 ff.) is a central part of the poem. Here we find an account of the specific tasks and the time of year when they should most advantageously be performed. The list is selective; thus, olive-growing is omitted. Nor is there a proper description of the physical frames surrounding agriculture, the farm, the village or the fields. Cattle-breeding is also left out; the most important livestock are oxen and mules as draught-animals. Hesiod's account may be supplemented by many glimpses of country life as found in the Homeric poems, especially those occurring in the similes. Their purpose is to illustrate a given situation in the narrative of the poems which often throw a very precise light on the individual activities described, but the interpretation of the more general position of agriculture in society, as presented in the poems, is more difficult. It is inherent in the genre to which the poems belong. The setting should be heroic, but at the same time intelligible to the audience. All in all, however, we may say that with these poets, who inaugurate what we call the historical period, we may form a reasonably good picture of agriculture as it was then, or at least aspects of it. After this, there is a long interlude. It is only with Xenophon's *Oeconomicus* that a coherent description of life and work in the country emerges through Socrates' discussion with Ischomachos. The dramatic dating of this part of the dialogue must belong to the period immediately preceding the Peloponnesian War, provided it makes sense to talk about a dramatic dating.[3] In the dialogue Socrates plays the part of the pupil, and Ischomachos emphasizes why agriculture is a natural phenomenon. Everyone understands that art immediately; but this cannot fool us. Knowing how to sow corn with the hand is one thing; having that special rhythm in your body which enables you to sow evenly over the entire plot of land tilled is a different matter. Xenophon doesn't master this side of the *techne* of agriculture from personal experience. He is the gentleman farmer who knows agriculture primarily as a spectator.

The third description of agriculture is given by Theophrastus, in *Historia Plantarum* and *De Causis Plantarum*. Here the experienced botanist reasons about cultivated plants, but the situation is viewed, as it were, from the point of view of the plants and not from the people who grew them. We find a number of interesting details, but a less coherent picture of agriculture as a whole. In addition, in many passages by the historians, the orators and not least by the comic playwrights we are given definite information about agriculture and glimpses of country life.

The state of the tradition at our disposal being as it is, conditions in Attica are mentioned first and foremost, and the general validity of the

---

3 Hanson 1983, 141.

information conveyed could therefore be made the subject of debate. Epigraphical sources often furnish us with precise information concerning isolated transactions such as leasing, proceeds by an auction and so on. They will be dealt with in the proper place. However valuable this may be, we must stress the difficulty in drawing conclusions from that which is specific as against that which applies generally.

The more traditional archaeological sources include excavated agricultural buildings (few of which, as we shall see, constitute reliable evidence), tools preserved, and pictorial representations of typical situations as they occurred in agricultural life. It is no wonder that vase paintings are dominated by picking of grapes and pressing of wine done by maenads and satyrs, but such representations, too, must of course be taken into consideration in their proper place.

Even by a combination of these very diverse testimonies a presentation will inevitably contain a series of lacunae. We shall not be able to arrive at a safe and uniform picture of agriculture, its produce and its productivity. We must endeavour to set forth the evidence for discussion and attempt to justify the individual links in the reconstruction. Perhaps the most prominent problem is the often very uneven geographic distribution of the sources. As a rule they refer to Attica, or else references are so vague and general that no precise location can be made. Hence it is imperative that a brief survey of the physical environment within which agriculture took place should be given.

1

THE GEOGRAPHICAL BACKGROUND

No sooner do we attempt to approach the topic of Greek agriculture than we are confronted with a question of principle: is 'Greek agriculture' the kind of agriculture practised by Greeks, or is it agriculture as practised in the area which has been called 'Greece' since Roman days? In the historical period Greeks were settled all along the coasts of the Mediterranean and the Black Sea. From the time of the Mycenaean civilization, and later on from the beginning of the historical period, the Greeks had increased the number of their settlements by the hundreds as if by gemmation; they had founded city-states wherever they could gain a foothold.

At this point we shall not pursue the age-old issue: was it the quest for land or the inclination towards trade that was the primary cause of this astonishing dissemination of Greek-speaking people?[4] We must assume, however, that from the beginning agriculture was the basic industry, and that the taking of land occurred approximately the way it did when the hero Nausithoos emigrated to Scheria: he built houses and sanctuaries, erected a city wall and distributed the arable land among the people (*Od.* 6.4 ff.). No matter where the Greeks settled, they brought with them their simple agrarian technology and endeavoured to establish themselves as much as possible in the way they were accustomed to at home.

Ideally speaking, it would be appropriate if we were able to present a detailed account of the entire Greek area and its natural resources, but such an attempt would take us far beyond the limits of this investigation. Instead, we shall emphasize general features and concentrate on Hellas proper, that is, the central part of present-day Greece with Attica, the Peloponnese and the islands, as well as the coast of Asia Minor, in other words, what in Archaic and Classical times might be called Greece of the city-states. Thessaly, Macedonia and Thrace constituted separate structures, but often with Greek colonies along the coast. If, occasionally, we turn our attention to places beyond this area, it will mostly depend on whatever source material has been preserved. We may assume that the colonists had been able to

---

4 Graham 1982, 157.

adjust to the new environment rather quickly. Otherwise they would not have been able to survive.

As far as climate is concerned, conditions in the Mediterranean zone are fairly even. The Mediterranean climate is characterized by its mild and rainy winters and dry and hot summers; but rainfall is unevenly distributed and frequently comes in the form of torrents during the winter months, which is the essential stage of growth of the plants. The summer is so dry that natural growth is brought to a standstill. However, this applies mainly to the lowland. In the mountainous areas the climate is rather Alpine: harsh winters with a great deal of snow and cold, and warm summers when the snow melts and provides the soil with large amounts of water. When the thaw comes, the rivers in the lowland swell, but dry up during the summer and leave the dried-out stony river beds with a very small amount of water, if any. This is well known to any tourist. The lack of water in the rivers in the hot summers means that there are seldom water resources for artificial irrigation during the summer months.

The contrast between mountains and plains has often been discussed by scholars. Few have described the Mediterranean world with greater insight than Fernand Braudel in *La Méditerranée et le monde méditerranéen à l'epoque de Philippe II* (1949). The chronological framework is early modern, but there are matters of general interest in this presentation which will undoubtedly stimulate the ancient historian. Historical geography has also contributed to our knowledge of natural resources. Here, we shall mention merely a couple of standard works on the topic, such as E.C. Semple, *The Geography of the Mediterranean Region: Its Relation to Ancient History* (1932) and M. Cary, *The Geographic Background of Greek and Roman History* (1949); as for Greece, the monumental work by Alfred Philippson, *Die griechischen Landschaften*, vols I–IV (1950–9) should be mentioned. Also useful is the Geographical Handbook series on Greece, published by the British Admiralty in three volumes (1944) and its American counterpart, the 'Army Service Forces Manual, Greece', available in a typewritten edition from 1943. Both of these practical handbooks contain a wealth of statistical information, largely based on facts assembled in *Annuaire statistique de la Grèce* (1933–9). Whatever these figures may lack in current interest, they gain strength in our investigation precisely because they hail from a period when very little industrialization of agriculture had taken place.

Generally speaking, we must accept that there are very considerable regional variations not only in the Mediterranean world, but also in Greece itself. The great formations of mountains stretching, broadly speaking, from north to south through Italy (the Appennines) as well as through Greece have been disrupted by later dislocations in the earth; in particular, central Greece shows us a multitude of different types of landscape, each of which is often of a very limited extent. This phenomenon has often been adduced as the geographical prerequisite for the particularism of the Greek city-states. Natural boundaries frequently set a limit to the small arable pockets of land that constituted the territory of the Greek city-state. Instead of entering into some sort of research concerning the soil conditions of individual areas, which would require a fairly long-winded tale, we shall confine ourselves to an important feature: the interrelation between elevation and rainfall. The orographic map of Greece shows the extent of the mountainous area, the Pindos Range with its elongation in central Greece, down to the Peloponnese, with a few massifs such as Mount Olympos, the highest point of which rises to 9,537 feet. If we combine this with a hydrographic (precipitation) map (Figures 1.1, 1.2), we find that western Greece has the highest amount of rainfall with the highest precipitation exactly in the Pindos Range. This is, of course, due to the fact that the moist autumn and winter winds that bring rains with them are predominantly western. Therefore, they discharge most of their rain while passing the mountains. The north winds dominate the summer months, and here again the mountains constitute a barrier forcing the air, as it rises, to discharge its waters. This orographical effect is important when we try to assess climatic variations in Greece, and, as the map shows, it spreads, with the result that the Cyclades are fairly arid, whereas the coast of Asia Minor and the nearby islands – Rhodes, Samos, Chios and Lesbos – are less so. It is noteworthy that Attica, the Islands and the Saronic Gulf appear to be the most arid zones in Greece, and it is this very area from which our literary sources are most abundant. So we find yet another limitation to the documentary value of our written evidence.

On the other hand, it would be dangerous to postulate that no climatic changes have taken place in the span of time between antiquity and today. It is well known that we have witnessed changes in climate also in historical periods, but every attempt to demonstrate that climate in ancient Greece was essentially different from that of our time has failed. We have to conclude that no methods are available to measure changes in climate and therefore accept, *a priori*, that climate is not likely to have changed radically. We shall have to accept the records from more recent periods and take them at face value. At the same time great uncertainty reigns when we come to the question whether the countryside itself may have remained unchanged. Needless to say, the countryside as such does not change its contours – at least not broadly speaking – but the way man has exploited his environment is a factor that constantly changes the landscape. Agriculture is based on exploitation of the soil – more or less efficient – and the land, therefore, has to be cleared of its natural growth of plants, primarily the forest. Thus the land is laid bare and exposed to erosion. The question of soil erosion, particularly in Greece where today the countryside is often dominated by naked rock, is and always has been a matter of interest, and more hotly discussed. Some scholars have attached importance to the overwhelming

THE ART OF AGRICULTURE

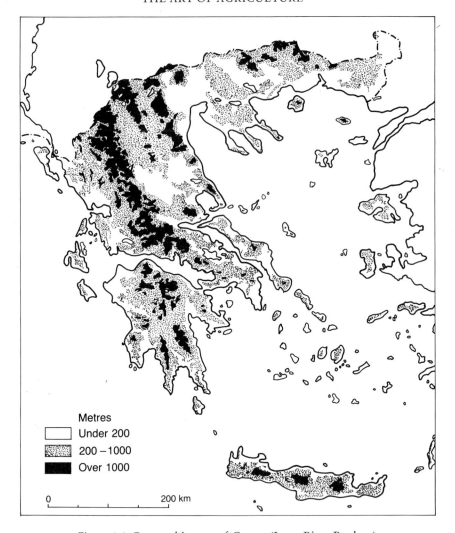

Figure 1.1 Orographic map of Greece (Inger Bjerg Poulsen)

destruction that deforestation and the grazing of sheep and goats have caused; others have referred to the inborn ability of the forest to regenerate as long as the soil is not exposed to constant tillage. With Plato's *Critias* (111 c) as a starting-point the pessimists claim that already in the Classical period central Greece was deforested; others would maintain that as recently as the latter centuries in modern times considerable areas of forest were still to be found in Attica: in any case, it wasn't until the fifth century that Attica needed to import wood in larger quantities, particularly from Macedonia, for ship-building. Analyses of pollen would have provided us with more

THE GEOGRAPHICAL BACKGROUND

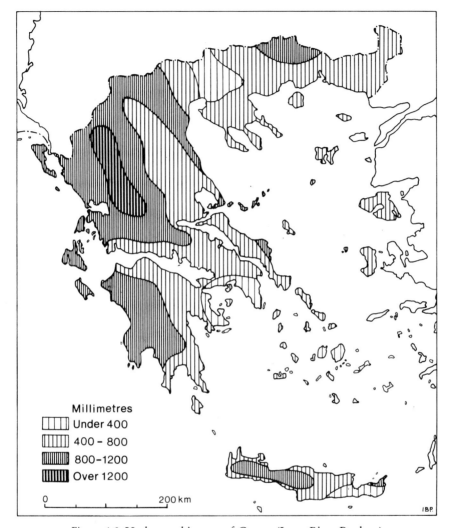

Figure 1.2 Hydrographic map of Greece (Inger Bjerg Poulsen)

reliable information on this crucial point, but inasmuch as pollen is preserved mainly in moist subsoil, such analyses are rare in Greece. The best-known examples are the investigations undertaken at Philippi in Macedonia and Lake Kopaïs in Boeotia. Both testify to a stable climate from the early Palaeolithic Age to our time, and to a considerable deforestation of the virgin oak forest during the Bronze Age. But let us bear in mind that this applies especially to the lowland. Erosion as a result of deforestation will hit the foothills of mountains, and from these zones we lack reliable data – investigations have not been undertaken and are not likely to be undertaken since pollen does not lend itself to examination in this particular soil.

The problem concerning the ancient forests has been dealt with recently by Russell Meiggs in *Trees and Timber in the Ancient World* (1982) with a concluding chapter on deforestation. The scientist, O. Rackham, has contributed an important study: 'Observations on the Historical Ecology of Boeotia' (1983). This is a botanical analysis of the growth of plants today in a limited part of Boeotia, comprising the plains as well the mountainous country, and an attempt to interpret their historical development, drawing on the botanical evidence, written sources and pollen analyses from Lake Kopaïs published previously. Rackham opposes the traditional concept of human activity as being destructive to nature. His interpretation of the sources leads on to the assumption that, by and large, the countryside and its vegetation have remained unchanged throughout the historical period, although, naturally, the flora is not in the least static. Deforestation is not in itself harmful and does not automatically lead to erosion; new vegetation, like maquis, gariga and steppe, replaces the forest together with the arable land. Many have looked upon these plant communities as inferior when compared with the forest, and many have regarded them as degenerate forest, hampered in growth and development owing to the constant overgrazing by sheep and goats. Rackham emphasizes their rapid growth, which prevents soil erosion just as well as the original forest. According to this interpretation the countryside is stable until the industrialization of agriculture makes radical changes feasible.

One may say that these results of botanical studies corroborate ecological points of view set forth previously, for instance by social anthropologists who have examined regional communities in Greece such as Hamish Forbes, Harold A. Koster and Nicolas Gavrielides.[5] They are, of course, only of limited geographical validity, but other investigations confirm the stability of the landscape and the limited erosion in historic periods. A similar constancy has been observed in the course of geological investigations in the southern part of Argolis.[6]

For anyone occupied with the history of agriculture in antiquity it is, of course, good news that the physical frames of agriculture are immediately comparable to what we witness today. However, this point of view must not be over-emphasized because it is scarcely true in its details, but only in its main features. It would not be wise to draw close parallels between conditions in antiquity and present conditions, nor would it be advisable to employ analogies, for instance of a quantitative nature. In the following pages we shall try to present a cautious interpretation of a landscape, and from among many candidates we have selected the Plain of Marathon. It has

5 In Dimen/Friedl 1976.
6 Van Andel/Runnels/Pope 1986, cf. Van Andel/Runnels 1987; Zangger in Wells, ed. 1992.

the advantage that it has been populated at least since early Helladic time and may, furthermore, presumably be familiar to most historians owing to its importance in political and military history. The so-called 'Marathonian Tetrapolis' appears to have been united with the rest of Attica relatively late since it continues to maintain its independent representation in Athenian legations at Delphi. Thus we may argue that under different circumstances the area might have developed into an independent city-state, but for reasons which we shall not discuss in this connection the area was absorbed into a larger entity.[7]

Geologically, the Marathonian Plain is an alluvial formation, approximately 10 km in length and 3 km wide, bounded towards the landward side by low limestone mountains which rise fairly steeply from the flat country and form valleys into the countryside. Through the most southerly of these flows the stream called Charadra, which bypasses the modern village of Marathona and flows into the Bay of Marathon, bounded to the north by the mountainous peninsula of Kynosoura. Until recently, the plain was characterized by vast swamps which, as is well known, play a considerable part in the reconstruction of the Marathon Battle; today they have mostly been drained and converted into arable land. Pausanias tells us (1.32.7) that a water-course from a shallow lake (*limne ta polla helodes*) discharged its waters into the Bay. Close to the lake it carried water which was well suited for grazing cattle, but close to the sea it was salty and contained an abundance of salt-water fish.

If one studies a picture of the area (Plate 1.1), approximately from the north to the south, one finds a clear image of the nature of the ground. The low mountains constitute a clear transition to the flat country. In the foreground you see the low wild flora in the stony ground. This flora also extends upwards on the mountains and consists, primarily, of evergreens such as bushes and grass. The herbaceous plants dry up in summer, but the evergreens provide a moderate fodder for grazing sheep and, in particular, for goats. The modern village has spread over some of the upper and dry part of the plain, whereas the lower and fairly flat part of the plain is arable land. The rather steep mountains offer little opportunity to cultivate the lower parts of the foothills by ploughing. Were they to be utilized, it would have to be by terrace farming (Plate 1.2) or by growing, for instance, olive trees which thrive quite well in poor and thin soil (Plate 1.1, left).

These plains have the advantage of being a marshy area with a natural growth of grass suitable for cattle. This was the case still at the time of Pausanias. In that respect, this typical landscape differs from most areas in Greece where natural growth of grass is rare in the summer months. So, in theory, we may assume that mixed farming in the proper sense of the word may have existed – a combination of agriculture and cattle-breeding – but

---

7 Pritchett 1960 offers an excellent description of the area.

Plate 1.1 The Marathonian Plain

Plate 1.2 Landscape with terraces, Methana

this is not the normal state of affairs in today's Greece, nor was it probably so in antiquity. The OEEC Report 56 (1951) which deals with *Pasture and Fodder Development in Mediterranean Countries* recommends fodder plants to be grown besides the other crops in order to produce farmyard manure. The report makes it clear that approximately 80 per cent of Greece must be regarded as mountain country, and states (p. 54) that 'the only climax grasslands are high mountain meadows with a wealth of grasses and forbs'. It is recommended that a series of cultured plants be imported: fodder plants, partly pulses, partly a series of wild-growing plants eventually to be domesticated, and alfalfa. At the same time, it is stated that the most serious difficulty connected with this reorganization of the management of farming is the fact that the individual farmer's land is too small for proper rotation of crops, ideally speaking with crops of one or several years for animal production. In 1951 animal husbandry had reached its nadir because of the civil war. Stock-breeding consisted mostly of draught-animals like oxen, donkeys and mules, together with a very limited number of sheep and goats which, during winter, grazed on fallow fields and in the summer months were fed on fodder that had been purchased or collected in orchards or vineyards where artificial watering had been used. Essentially, larger livestock was transhumant – that is, summer grazing in the mountains and winter grazing in stubble and fallow fields as are found in the lowland, and, of course, on the untilled areas on the lower foothills encircling the Marathonian Plain. This transhumance or, as it is less appropriately called in the report, 'nomad farming', was considered very harmful to forest and natural flora with harmful side-effects like erosion. In consideration of the recent studies mentioned above, we may be permitted to disregard, to some extent, the feeling of impending disaster that is characteristic of this and other descriptions, particularly with regard to the destructive effect on the countryside for which the goats have been blamed.

The question has often been discussed whether ancient agriculture had a similar dividing-line in agriculture between a minimum of stock-breeding and transhumant cattle-rearing which utilized the natural summer resources in the mountains. Later on we shall revert to this problem, which is of fundamental importance for our understanding of what the soil could yield. Without a very considerable supply of organic material, the soil must have longish rest-periods so as not to be exhausted. Fertilizer, which is used nowadays, was unknown in antiquity; therefore, the alternative to effective mixed farming must be rather long periods of leaving the land fallow. So on a yearly basis the area that can be effectively tilled is drastically reduced.

Naturally, the Marathonian Plain is merely one example and certainly not typical, partly owing to the abundance of water. It would be difficult to find any landscape that would be typical for Greece as a whole, and all scholars agree in emphasizing the differences as being typical.⁸ We are faced not only with variations between the territories of the city-states, but also with variations within very small areas which makes any kind of generalization difficult. Investigations of these differences within a modern local community have made Hamish Forbes⁹ emphasize the advantage of dispersing the lands of the individual farm: it minimizes the risk of crop failure. The peninsula Methana on the north coast of the Argive peninsula is admittedly characterized by enormous differences (that is, when we consider heights above sea level), but the tendency towards dispersal of one's land may be observed in many other places, so that it may be described as typical of non-industrialized farming in Greece. We shall revert to the question of whether, and to which extent, similar tendencies were prevalent in antiquity. Among other considerations, it depends on the complicated rules regulating inheritance and dowries. A bold attempt at emphasizing differences has been made by Robin Osborne in *Classical Landscape with Figures: The Ancient Greek City and its Countryside* (1987). Here the author undertakes to examine a number of different city-states and relates the structure of settlement to the natural possibilities offered by the countryside with regard to agriculture, husbandry, mining, quarries and so on. The book often presents surprising and convincing solutions to problems, and much of what will be discussed in the present studies is dealt with in his book. Whereas it is Osborne's intention to underline the differences between the city-states, we shall attempt to concentrate on the more general feature that, in spite of all differences, seems to mark the main industry in Greece, namely, agriculture.¹⁰

---

8 Hammond 1963 gives a very short and useful description of the Greek landscapes.
9 1975b.
10 In May 1990 a small symposium took place in Athens at the Swedish Institute. The topic was 'Agriculture in Ancient Greece' and the papers are to be published by the Swedish Institute by Wells in 1992. As far as possible, references to these papers will be made in what follows.

2

TILLING AND CROPS

Methods of tilling in Greek agriculture have frequently been discussed in more recent literature. The comprehensive articles in *Pauly–Wissowa–Kroll's Realencyclopädie* by Olck, Orth, Kornemann and others are still of fundamental value. There we find references to earlier studies and in particular to the written sources. Separate chapters on tilling may also be found in the classic presentations by Gustave Glotz (1920), Paul Cloché (1931) and R.J. Hopper (1979); however, during the past generation a very far-reaching perspective has been added to the history of agriculture, partly by Ventris's decipherment of the Linear-B tablets from the Bronze Age and partly by virtue of the great expansion of Bronze Age archaeology where the study of Mycenaean civilization is a separate and rapidly growing field.

In this chapter we do not intend to pursue the agrarian history of the older periods, but with our new knowledge we find ourselves in a new situation when trying to describe the agrarian conditions in Archaic and Classical Greece. Here agriculture rests upon an immensely long tradition which takes us back in time for thousands of years. It is a particular problem whether the ancient agrarian community lived through a radical change during the so-called 'dark' centuries which followed the collapse of the Mycenaean Palace Culture about 1200 BC until the phenomenon of the rise of the city-states occurred, dated to the eighth century. Of the greatest importance for historical research is the fact that the art of writing seems to have been lost so that the notions that Greeks of later periods entertained concerning their own past were very vague. A. Snodgrass (1977) maintains – together with others – that we are dealing with an essential limitation of farming in favour of extensive cattle-breeding, in other words, from a certain evolutionist point of view, a cultural step backwards. Other scholars adopt a more sceptical view towards the hypothesis of a transition to cattle-breeding, as well as towards the interpretation of a phenomenon like that as being a backward step.[11] It is, however, still an open question whether the

---

11 Snodgrass 1977 maintained that there was a drastic decline in the population in the post-Mycenaean time and an even more drastic increase of perhaps 4 per cent yearly during the geometric period. The latter is challenged by Hansen 1988, 9.

collapse of the Mycenaean culture as an economic system coincided with the collapse of the agricultural system. One may well suppose that the peasants who were presumably dependent on the palace continued to till their land during the so-called 'dark' centuries as they used to, but without having to deliver a surplus to the princes in the palaces. Along with this discussion follows the complex of problems which has been called the 'Dorian Invasion', originally a theory developed to account for the spreading of the Greek dialects and usually combined with the ancient tradition of the return of the Heraklidai. We shall not enter into a discussion of this theory but choose our starting-point in the historical situation as it was at the time of the introduction of the alphabet in the eighth century, however unclear it is. The only thing that is certain is that the growing of what Colin Renfrew has called the Mediterranean triad – grain, vine and olives – at this time had a history of a thousand years behind it. As mentioned above (p. 7), this is not the immediate impression you get when reading the two great poets who stand at the dawn of this period, Homer and Hesiod. The former is acquainted with olive-oil, but only once does he refer to the growing of the olive tree (*Il.* 17.53 ff.), whereas the latter does not at all mention this cultivated plant and the use to which it was put. This led Victor Hehn (1870) to assume that the growing of the olive tree was a fairly new phenomenon at the time when the poems were composed, but this is amply contradicted by archaeological and epigraphic evidence from the Mycenaean period. Therefore, we can draw no conclusions *e silentio* when we consult these authors.

Agriculture in Greek epic literature has been dealt with briefly by Frank H. Stubbings (1963) and more comprehensively by W. Richter (1968). Both aim at throwing light on the realia of the poems, which from the point of view of agrarian history may be described as a restriction. This seems to be the case especially in Richter's treatment: he supplements our knowledge partly by introducing the earlier Mycenaean evidence and partly later material. In this way it can be concealed how much or how little of our knowledge does in fact rest on contemporary sources. Here we shall choose a more systematic treatment of the three main crops in the Mediterranean triad and leave the chronological principle in the background. The reason for this is evident. A strictly chronological treatment of the phenomena as they are described in the sources could convey a false impression that, by themselves, the sources aimed at completeness whilst in reality they have maintained a selective attitude towards the subject regardless whether, as with Homer, it is essentially an illustration or, as with Hesiod, a natural vehicle for basic moral deliberation.

GRAIN

It is well known that the word 'grain' is the designation for a series of grasses which, apparently over a very long period, were domesticated in what is called the transition to the Neolithic Age. Some have described this process as the most far-reaching revolution in the history of mankind because the growing of grain made permanent residence possible.[12]

The Greek word which, in the Classical period, most closely corresponds to 'grain' is *sitos*. But Homer applies it not to grain but to food made from grain, or bread, as opposed to, for example, meat. The species of grain mentioned by Homer are *pyros*, *zeia* and *krithe*, all of which are mentioned together in *Od.* 4.604, *pyros* and *krithe* only in *Od.* 9.110 and 19.112. The names of the species of grain also appear in decorative epithets, for instance *polypyros*, *zeidoros* and others, frequently associated with a specific area.

It is odd that Hesiod, in his *Works and Days*, does not mention the species of grain by name, although he does use epithets like *zeidoros* (*Works* 173). In his *Oeconomicus* Xenophon refers to grains like *pyros* and *krithe* (16.9), but later on he uses the word *sitos* as a major term (17.6), cautiously translated as 'food' by Marchant in the Loeb edition. It can scarcely be doubted that in this context Xenophon thinks of grain in general. The use of the word *sitophilos* concerning merchants would also indicate the meaning of the word as the major term, like the name of the college of office-holders *sitophylakes*, and the term *sitopoles* in Lysias 22.

In botanical terminology, as used by Theophrastus, *sitos* and the adjective *sitodes* clearly designate grain as against, for example, leguminous crops, but it is noteworthy that millet (*kegchros*), to choose an example, is not looked upon as belonging to the types of grain in a closer sense (see for instance *Historia Plantarum* 8.1.1). This is an excellent example of how the ancient scientist applies other criteria for his systematization than contemporary scientists do.

Botanical identification of ancient plants is, of course, always a problem, but there can be no doubt that *zeia* and *pyros* are species of wheat (*triticum*), the former emmer wheat (*triticum dicoccum*) and the latter wheat proper (*triticum vulgare* or *cereale*), whereas *krithe* is barley (*hordeum*). It does, however, remain a question how much reliable information concerning ancient grain has been thus conveyed, inasmuch as the species of grain are in a constant state of development, particularly in our own time when it has become possible to create new variants with specific genes. Still, we must assume that at an early time a certain stability must have been arrived at, although it will be difficult to say precisely when. It is noteworthy that not many essential archaeological data concerning grain and other cultivated plants from the historical period appear to have been published, whereas

---

12 Fundamental is Jardé 1979, Heichelheim 1935 and Moritz 1955a and 1955b; most recently there is Amouretti 1986, more systematic than historical.

prehistoric archaeology has been aware, to a far larger degree, of the possibilities of studying organic material, for instance, carbonized grain or impressions in pottery. In fact, we know more of cultivated plants in their earliest days than in later periods. We can conclude that experiments undertaken by comparing carbonized grains from an early period and corresponding grains from species of corn now in existence testify to some increase in the size, but the very fact that we can identify prehistoric species by comparing them with modern species shows that we are dealing with a reasonable degree of constancy. In her book *Palaeoethnobotany. The Prehistoric Foodplants of the Near East and Europe* (1973), Jane Renfrew discusses a number of these problems in her introduction and conveys a series of results in tabular form. Therefore we can refer to this book in general as far as the earliest history of the species of grain and of most other cultivated plants is concerned.

Pictures of ears of corn from Archaic and Classical times in Greece are by no means rare. We have ears of corn on vases (Plate 2.1) as well as on coins (Plate 2.2), but here of course the artistic tradition plays a decisive role. Does the artist depict the existing types of corn or single grains with minute precision, or do other considerations (such as the decorative) influence him? Without definite traces of the grain itself this cannot be ascertained. We do, however, get an impression of the ear of corn, and it is, for instance, possible to distinguish between two-rowed and six-rowed barley on coins. On the other hand, a more accurate use of the pictorial representations is scarcely possible. The same applies to the interpretation of the variations within the individual species. Theophrastus indicates a number of these, but a closer identification is hardly feasible. At this point we shall restrict ourselves by referring to the comprehensive Index of plants in the Loeb edition of *Historia Plantarum*.

The written sources are in agreement that autumn sowing is normal and optimal. Hesiod introduces his agrarian calendar by combining the most important tasks of the farmer, sowing and harvesting, at the setting and rise of the Pleiades, that is, November and May (*Works* ll. 383 ff.) and more specifically indicates the autumn flights of the cranes as a signal for the beginning of autumnal work (ll. 448 ff.). Sowing takes place after ploughing, according to Hesiod after the third ploughing (ll. 462 ff.), whereupon sowing is done by hand. A boy hoes the seed into the ground lest the birds eat it. Clearly, we are dealing with the sowing of a field that has lain fallow for a year (*neios*); that is to say, the poet presupposes a two-field system with one year's crop followed by one year's fallowing. Homer also describes ploughing in several passages, as a rule of the fallow field; thus for instance the description of the shield of Achilles (*Il.* 18.541 ff.), the similes (*Od.* 13.31 and *Il.* 13.702 ff.) and the important place in Dolonia (*Il.* 10.351) where mules are used to work the plough. As an epithet the adjective *tripolos* ('thrice-ploughed') is used several times (*Il.* 18.542; *Od.* 5.127, cf. Hesiod,

Plate 2.1 Triptolemus with an ear of corn, Attic red-figure hydria, Ny Carlsberg Glyptothek, Copenhagen

Plate 2.2 Ear of corn, stater from Metapontum, National Museum, Copenhagen

Theogony 971). It seems confirmed that it was considered normal to plough the fallow field three times. One might also adduce the name Triptolemos, the ancient hero who is associated with the Demeter festivals in Eleusis, as a testimony for the great age of the biennial system.

Xenophon too apparently counts with alternating corn crops and fallowing (*Oec.* 16.10). The ground is loosened in the spring before grass and weeds have seeded so that by ploughing down before the spreading of seeds they can serve as green manure. The sun dries and makes the soil friable in the course of the summer when the earth should be ploughed as often as possible (*hoti pleistakis*). At the beginning of the rainy season in the autumn sowing with the most plentiful grain takes place in the rich soil, less so in inferior, but there is no reference to hoeing in of the seed. In several places Theophrastus mentions sowing in connection with a variety of cultivated plants, most clearly *De Causis Plantarum* 3.20.1. Here he distinguishes between the treatment of light and heavy types of soil, and digging or hoeing (*skaptein*) is mentioned as an alternative to ploughing. Light soils are dried out too much by ploughing in summer whereas they benefit from ploughing in winter; conversely, heavy and moist soil benefits from summer ploughing. As far as the time of sowing is concerned, Theophrastus refers to Hesiod's determination of the setting of the Pleiades, and he recommends the sowing of barley before the wheats (*HP* 8.6.1). It is precisely in this context that you see the difficulties in using Theophrastus as a source in the history of agriculture. He treats the sowing of cultivated plants together, and it is often difficult to tell from his succinct choice of words exactly what he has in mind. Numerous cultivated plants are mentioned but no priority or any estimate of profits is given. The question of the possible development towards a more complicated rotation of crops will be discussed in connection with the other cultivated plants.

Apart from winter crops, sowing in spring is also attested. Hesiod (*Works* ll. 485 ff.) seems inclined to regard spring crops as an emergency in case you missed winter sowing, whereas Theophrastus mentions *dimenoi* and *trimenoi*, crops which ripen in two or three months (cf. *HP* 8.4.4 and *CP* 4.11.4). However, the latter do not appear to have been able to compete with the winter crop.

It must also be remembered that we possess a series of representations of ploughing and sowing in contemporary art. These will be dealt with in the following chapter on implements. Furthermore, inscriptions provide us with information: for instance there is a famous document of leasing from Amorgos (*SIG*³ 963) and a somewhat obscure inscription from Piraeus (*SIG*³ 965), where the tenant is allowed to plough as he sees fit for the first nine years whereas in the last year he is allowed to cultivate only one-half so that the other half shall remain fallow out of consideration for a tenant who might take over. To take this as evidence for a general use of a more complicated rotation is a far step.

Whereas Hesiod is not interested in the field between sowing and harvesting, Xenophon recommends hoeing of the cornfield if violent rainfall has laid bare the roots or covered the tender seed-leaf with mud, just as weeding should be undertaken (*Oec.* 17.12 ff.). Theophrastus briefly mentions two jobs in the field, *skalsis* and *poasmos* (*CP* 3.20.6, cf. 4.13.3), and we must assume that this is the designation for weeding similar to that given by Xenophon, but here we are not dealing with the job description but with an illustration by example. Weeding of fields has been known all the way to modern times when chemical control of weeds has prevailed. As mentioned, the time for harvesting is given by Hesiod to the rise of the Pleiades (see also *Works* ll. 571 ff.). It is done by means of the sickle (*arpe*), and the corn is brought home. Here it is threshed on the threshing floor (ll. 597 ff.), and the grains are stamped free. Still later, straw and chaff are stored in barns, upon which follows a period of rest.[13]

Harvesting and threshing have also been described by Homer, first and foremost on the shield of Achilles (*Il.* 18.550 ff.) with a very vivid description of reaping as well as sheaf-binding, but also in a simile (*Il.* 11.67 ff.). For threshing see *Il.* 5.499 ff. and 20.495 ff.

Xenophon gives us a fairly detailed description of the mowing as well as of the work connected with harvesting and threshing (*Oec.* 18.1 ff.). The reaper stands with his back to the wind. If the straw is short, it is cut at the root; if it is long, it is taken at the middle so as not to make the threshing difficult. The tall stubble can then be burnt or brought to the dunghill or to the compost heap; in other words, here we must presuppose a particular mowing of the straw. As in Homer, threshing takes place by driving the draught-animals round the threshing floor, and the ears of corn are thrown under their hoofs, whereby the kernels are trodden free, and retained by winnowing. The passage is an excellent demonstration of Socrates as a star pupil who will quickly come up with the right answers.

As a botanist, Theophrastus is not particularly interested in the harvest. He notes that barley stands on its root for seven or eight months, wheat somewhat longer (*HP* 8.2.7). This means that harvest occurs in May or June, but nothing is said as to how it takes place. It probably cannot be demon-

---

13 Gallant 1982 has a bold and wrong interpretation of this verse, see esp. p. 114. The words *chortos* and *syrphetos* are translated into 'litter and fodder' – perhaps by *hysteron proteron* – and he continues: 'this litter is presumably the stubble from the grain fields; the fodder on the other hand may well be crops such as vetches, lentils and lupins planted on the land that was ploughed in spring.' So he has introduced a three-field system with rotation, but the plants suggested are not to be found in the poem. The description follows the threshing on the threshing-floor near the house. 'Stubble' therefore as a translation of *syrphetos* gives no meaning; it is left on the field. The word means 'anything dragged or swept together' (Liddell, Scott and Jones), that is, probably, the 'chaff'. *Chortos* normally means 'grass for fodder'. Xenophon (*Anabasis*, 1.5.10) has *chortos kouphos* ('hay') put into leather sacks to cross the Euphrates. In our context it could, of course, be hay; but rather it is the straw, as by threshing you get three products – grain, straw and chaff. For *chortos* in Egypt see Schnebel 1925, 211 ff.

strated more clearly that Theophrastus is not interested in agriculture as such, but only in its plants. As far as the size of the crop is concerned, the ancient sources fail us completely. Naturally, to many historians this is the most important issue, but neither yield per land-unit nor seed-to-yield ratio is known, and that is the very reason why we may guess. Pessimists like Jardé are constantly thinking of non-industrialized modern Greece, whereas others are more optimistic. In a controversial article Garnsey demonstrates clearly that the crops of Attica can be calculated only as a purely speculative arithmetical example. Therefore, we shall refrain from entering into this kind of guesswork, but confine ourselves to a reference to Osborne, who also discusses this matter and cautiously points to three- and tenfold as possible extremes for the seed-to-yield ratio. At the same time he emphasizes the fact that there must be great variations depending on soil and precipitation, and it is a question whether we are entitled to speak of a 'normal crop'. From the point of the farmer the most important question of all will be his ability to survive during lean years also.[14]

WINE

Whereas the growing of grain is a constituent element in Hesiod's agrarian calendar from the time of sowing in the autumn till harvest, then threshing and renewed sowing, the same does not apply to perennial crops like grapes, fruit and olives.[15] The pruning of the vine is mentioned briefly as work that must be terminated before the coming of spring, sixty days after winter solstice at the rising of Arcturus at dusk and the coming of the swallow (*Works* ll. 564 ff.). The wine harvest takes place at the heliacal rising of Arcturus (ll. 609 ff.). The grapes are dried in the sun for ten days, covered for five, and on the sixth day 'the gifts from highly generous Dionysos' are poured into vats (*aggea*). Here, nothing is said about pressing.

In Homer's description of the shield of Achilles there is an account of a well-planned vineyard, surrounded by a ditch and a fence. Along the road to this vineyard young men and women move together in flocks in order to pick the grapes and put them in wicker baskets while a boy plays an instrument and sings (*Il.* 18.561 ff.). The garden of Alkinoos (*Od.* 7.112 ff.) also includes a vineyard, facing south. Some grapes are laid out to dry, others are being picked and some are being trampled. At the same time there are vines in bloom while others carry ripening grapes. In this way the poet succeeds in conveying a paratactical description of the vineyard at different times of year, clearly an obvious counterpart to a dull everyday life which forms the frame of Hesiod's poem. The Scheria of Alkinoos is, of course, a Greek land of happiness, and we are not told much about the labour

---

14 Garnsey 1985 and 1988a, 89 ff., Osborne 1987, esp. pp. 44 ff.
15 Often Greek and Roman vine-farming are treated together: Billiard 1913, Jardé 1957, Hanson in Wells, ed., 1992.

involved. Both poets describe the vineyard as something already in existence and well organized in contradistinction to the description of the island of the Cyclopes. There everything grows by itself, thus also the vine, a gift the blessings of which are unknown to the barbarian giants (*Od.* 9.106 ff.). Only old Laertes is seen digging round a tree (*Od.* 24.226); apart from that, interest is centred round the fruit and to all that is well organized, none to the work itself.

Very different is Xenophon's discussion of the planting process. Once Socrates has realized, with wonder, that he does in fact know everything concerning the growing of grain, he and Ischomachos tackle the planting of trees (*Oec.* 19.1 ff.), but it is only at a later point (19.12) that it becomes apparent that they are dealing particularly with the planting of the vine. The starting-point is the size and depth of the hole designed for the plant, seen in relation to the natural moisture of the soil. In dry earth the hole should be deeper than in moist soil, and the bottom should be covered by well-tilled earth.

However, the peculiarity in this description is that Xenophon seems to think that the cuttings should be planted direct on the habitat. The terminology is not entirely clear; 19.4 refers to *ta phyta*, 19.8 refers to *blastos tou klematos* and 19.9 to *klema*. In the latter case it is clear that we are dealing with a cutting because it should be placed in a slanting position in the ground in order to take root, and when it is recommended that it be planted firmly, it is lest the cutting rot if the soil becomes too moist. We have now reached a strange point in our written tradition concerning vegetative propagation of trees in ancient Greece. Today we know various forms of this method of multiplying trees:[16]

1. cutting, whereby you lop off a scion, plant it and wait for it to take root whereupon it can be replanted in its own habitat;
2. suckers with offshoots, where the parent plant itself forms a new plant from its roots. This new plant has its own system of roots and can therefore be transplanted;
3. layering, where a twig is bent down and covered with earth after which it will take root and can be transplanted; and
4. grafting, where a scion is 'planted' in another tree.

The essential point is that the cutting does not take root until separated from the parent plant, and this must take place before the leaves of the cutting have unfolded. Normally cuttings are planted in a specially well-prepared soil that can be watered intensively because the cutting does not endure desiccation. The Roman agronomists are fully aware of this and describe very convincingly the grounding and care of a nursery (*seminarium*), but Xenophon indicates neither the time of year nor the particular place and

---

16 Hartmann/Kester 1968 have an excellent treatment of the subject.

particular care that is essential for success. Another detail that is not entirely clear is the question whether Xenophon believes that the cuttings should be placed one by one in a hole or that possibly a series of cuttings should be planted in a trench (as apparently Marchant thinks, inasmuch as he renders the Greek *bothynos* by 'trench', in his translation of *Oec.* 19.3). The word occurs but rarely, and what may account for the meaning 'furrow' is, of course, that depth and width are indicated, but not the length of the furrow. A little later the word *bothros* (19.7) is used; this is normally the designation for a hole. This may seem to be of little consequence, but it is a question whether we are dealing with individual planting at the final habitat, or with a somewhat more artificial production of cuttings in greater quantity, perhaps placed in a long row with a view to later transplantation. Nothing in Xenophon's text would indicate the latter.

As a botanist, Theophrastus naturally takes an interest in the gemmation of plants. The propagation of trees is dealt with in the second book of *Historia Plantarum* as well as in the third book of *De Causis Plantarum*. In both cases we are faced with a generalizing description with examples taken from individual trees to illustrate what is meant. Only rarely do we find a proper description of the work connected herewith. One exception, however, is *CP* 3.11 ff., which provides us with a more thorough account of viticulture.

The second book of *Historia Plantarum* begins with an enumeration of eight different ways in which plants propagate their species. Out of these, five are vegetative and brought about by human interference. Unfortunately, Theophrastus takes it for granted that the reader is fully familiar with the meaning of the terms, and therefore provides us with no explanation or definition of the words employed. Presumably, he picks out colloquial words, but elsewhere he admits to the inherent difficulties in distinguishing between the precise meaning of the terms used, particularly when dealing with twigs or branches used for grafting and cuttings (*CP* 5.1.3).

One might have expected him to continue using the terminology introduced at the beginning, but this is not the case. Elsewhere he is fond of using words the meaning of which is not entirely clear to us. Hence, once we get down to details, there is a series of problems where interpretation is extremely difficult or even impossible.

One difficulty is that the verb *phyteuein* (as well as related concepts – *phyteia*, and so on) is used about the placing of cuttings as well as about transplantation, layering, grafting and budding. This difference may be of less interest to the botanist who takes an interest in plants and their growth, but it is admittedly difficult for the historian interested in the development of agriculture. In each individual case one must consider which operation is in fact involved. In the following we shall attempt to demonstrate some of the difficulties that arise when you use Theophrastus as your source. A natural starting-point will be the account of cultivated trees as found in the third book of *De Causis Plantarum*, but also taking into account the second book of *Historia Plantarum*.

The time of planting is discussed in some detail in both contexts, autumn planting as well as spring planting (*CP* 3.2.6 ff.). The hole prepared for planting is dealt with in both places, using the word *gyros* as a technical term; that is to say, it is a (round?) hole, not a trench or ditch, although Wimmer prefers to render it by *scrobes* whereby, perhaps, he leads his reader to believe that we are dealing with the mass production of transplantable trees as we know them from Roman authors (*CP* 3.4; cf. *HP* 2.5). The holes should be dug well ahead of time because summer sun as well as winter cold makes the earth brittle. In *De Causis Plantarum*, it seems, the cuttings are foremost in the author's mind since the concept of taking root is a central topic (*ritsosis*). In the sequel mention is made of shoot which should preferably be taken for cuttings and scions, but in this context none of the words from the primary description in *HP* 2.1 is used, but rather terms like *phyteuterion* and the verb *moscheuein*. A little later (5.3) the term *promoscheuein* is used, apparently intended to denote layering but still in a very general context with no reference to any particular tree. Hence, it is very difficult to include the information obtained in a more specific description of the process. We are not in a position to see the different phases of the vegetative propagation clearly, and Theophrastus has not succeeded in establishing an unambiguous terminology by which he might have told us what is in his thoughts.

In spite of these difficulties, Theophrastus remains the main source of our knowledge concerning Greek viticulture. Reproduction by seeds does not occur in connection with cultivated trees because trees that stem from sowing will often revert to a wild form (*HP* 2.2.4), and here the vine is specifically mentioned. Cuttings are widely used (Plate 2.3). Grafting does not appear to have been used (*HP* 2.5.3), and the question of layering can scarcely be solved with any degree of certainty owing to terminological difficulties, although the assumption would seem reasonable enough. The vine is grown both as a separate stem as well as climbing upon a different tree (*anadendras*, *CP* 3.10.8), and we are dealing with an enormous wealth of species where the grapes will acquire taste and character depending on the soil and placing in relation to the sun as well as precipitation, the vine requiring much moisture.

Once the vineyard has been laid out, it requires a great deal of tending throughout the year. Here, fortunately, we are in a better position because this is the type of work that Theophrastus analyses in his systematic description of viticulture (*CP* 3.11 ff.).

The starting-point is represented by the choice with which one is faced concerning the various sorts and the importance of the soil. The description is kept on a theoretical level, and the sorts are not itemized by name, nor is anything said about their yielding capacity or their profitableness. Interest is concentrated on the xylem of the vine. Reflections on the laying-out of new

Plate 2.3 Cuttings of vine, Ithaka

Plate 2.4 Vineyard just before pruning in January, Methana

Plate 2.5 Vine, Chios

vineyards follow, also with regard to the plant hole, *gyros*. Digging at the beginning of spring is recommended. Pruning of the newly established vineyard is equally important, inasmuch as a vigorous dehorning will strengthen the roots. It can be undertaken immediately following the setting of the Pleiades, that is at the end of October, but in many cases the pruning should be delayed until the third year.

Once the grape vine has developed fully, the annual pruning is one of the most important tasks to be performed (Plate 2.4). Botanical reflections on the nature and condition of the xylem in relation to the soil are still the essential concern (*CP* 3.13). In a dry and hot climate the pruning takes place directly the leaves have fallen, whereas in cooler and more humid areas it is postponed until the new buds have developed; in *HP* 3.5.4 this is connected with the rising of Sirius and Arcturus. Barley or beans may also be sown between the vines in good soil so as to hamper the growth of the vine – in other words, a technical reason for the – undoubtedly – wide-spread practice of intercultivation which, of course, has reasons of an economic nature as well. The staking of the vine is closely connected with the cutting. Here, Theophrastus is not very informative, and we hear nothing about the stakes, trellis-works and similar features that play a considerable role in Roman technical literature. Especially when dealing with intercultivation we might have expected to hear about the vine having been bound up, but there is no mention of it. Often, like today, the vine may have been grown on dwarf stock without the necessity of binding, but supporting poles for the vine are mentioned elsewhere. In the splendid description of the vineyard in *The Shield* by Hesiod (l. 299), the sticks (*chamakes*) are made from silver, and the supporting poles (*charakes*) are mentioned also by Aristophanes. We do not know to what extent the binding of the vine was a normal procedure.

Thinning the shoots and leaves, *blastologia*, is likewise important. The former takes place as soon as the clusters of blossom appear, the latter immediately before the bloom. Between these events, the second tilling of the ground takes place. The nipping that serves the purpose of removing superfluous buds and leaves so that the vine may concentrate on developing its fruits is described, but with nothing like the precision and wealth of detail that we find in the writings of Columella, the wine expert, centuries later.

Finally, while the grapes are ripening, dust may be raised by hoeing the soil so that a thin layer of dust is brought to cover the grapes (the so-called *hypokonisis*). The effect is undoubtedly a delay in the ripening of the grapes at the time when ripening would otherwise have occurred, thereby increasing the sugar-content. There follows a discussion which amounts to the question whether the grapes are invigorated by the thin layer of earth. It is characteristic that the running description of the vine should end here, and once more we may note that the botanist takes no interest in the harvest as such.

Besides cutting and pruning, tilling the soil is the major task in the vineyard. This has been described previously by Theophrastus in the general part of his account, that which also applies to cultivated trees (*CP* 3.10). Here, too, he discusses the cutting of the roots and the use of manure which, in his opinion as far as the vine is concerned, should be limited to every fourth year or at even longer intervals (*CP* 3.9.5). He is fully conscious of the importance of airing the soil and of weed-control undertaken by tilling, as well as realizing that pruning the root forces it downwards, thereby making it more resistant in a period of drought.

With the information concerning viticulture, a picture takes shape showing a great degree of care and tending. Whereas the earlier sources are selective, Theophrastus gives us the first consistent description of the continuous processes from pruning in winter, cutting of roots and digging, until the concluding work prior to the ripening of the grapes. The actual vintage, however, is mentioned only by Homer and Hesiod, but frequently shown in art together with the stamping of the grapes. We shall revert to this subject in Chapter 3.

The problem of the scope and increasing professionalism of propagation is important when one weighs the possibilities inherent in agriculture with regard to increase of production and re-organization. We shall bear this in mind when dealing with the last element in the Mediterranean triad.

OLIVES

By and large scholars agree on the question of the wild forebears of grain and vine that can still be traced. There is less agreement with regard to the relationship between the improved olive tree (*olea europaea*) and the wild tree (*olea sylvestris*) found everywhere in the Mediterranean area.[17] Jane Renfrew (1973) and others consider the *olea sylvestris* a wild form of the cultivated tree. The wild form is a small, slightly thorny bush-like growth with rather small and more rounded leaves, whereas the improved tree is characteristic for its long and slender leaves and its heavy, gnarled trunk (Plates 2.6, 2.7). As the authentic ancestor of both, Renfrew points to the so-called *olea chrysophylla* and its variants. Others are of the opinion that the so-called 'wild' olive tree is the ancestor of the cultivated one. However that may be, we must conclude that a precise dating of the introduction of the cultivated variant cannot be given, inasmuch as the *olea sylvestris* also produces fruits rich in oil, but with much smaller pits and fruit pulp and consequently having much less oil-content. It does seem clear that the cultivation of the domesticated form, *olea europaea*, is firmly established in Greece at least from the time of the palace cultures of the Bronze Age.

As mentioned above, Homer and Hesiod furnish no information with

---

17 Pease 1937b, Amouretti 1986, further Forbes/Foxhall 1978, and various articles concerning modern cultivation in Diemen/Friedl 1976.

Plate 2.6 Wild olive, Athens, Areopagus

Plate 2.7 Domesticated olive, Methana

regard to cultivation of the olive tree.[18] Xenophon goes into some detail concerning its planting (*Oec.* 19.13). The plant hole has to be deeper than when planting the vine, and with the cutting follows a small piece of older wood so that it is not merely the fresh shoot but a mallet or heel cutting that is planted. It is mentioned that you look at the plant holes along the roads. The cutting is covered with a small amount of clay and a potsherd, presumably in order to prevent dehydration. The passage is so brief that it is difficult to arrive at a clear understanding, but it can be supplemented by comparison with Theophrastus; unfortunately, the latter does not offer a systematic description of the tending and care of the tree as was the case concerning the vine. One has to piece together information gleaned from the more general portions of his writings. In the introduction to the section on the propagation of trees (*HP* 2.1) he informs us that the olive tree is reproduced vegetatively in all the ways mentioned except as far as cuttings of completely fresh top shoots are concerned. Later (2.5.4) he mentions root-suckers and cuttings with a small piece of xylem (*hypopremna*),[19] just as one can plant a branch which has been split at the lower end, placing a stone on top, likewise presumably against dehydration. Like Xenophon, Theophrastus refers to the practice of covering the uppermost cuts of the shoots with clay and a potsherd (*CP* 3.5.5). In the large section on grafting and budding (*CP* 1.6.10) it is mentioned that both procedures are employed in the cultivation of olives, the wild olive tree (*kotinos*) being used as the stock.

Thus, we know a great deal about the possibilities inherent in the vegetative propagation of the tree, but we do not know which procedure was typically employed by the Greek peasants. Theophrastus catalogues and exemplifies, but it is still the botanist speaking. He is also able to mention peculiarities like, for instance, implements made of olive wood, a door stop and an oar having taken root (*HP* 5.9.8), but it is of no great help to the historian who deals with agriculture beyond the point that it emphasizes the incredible ability of the tree to reproduce itself even under unfavourable conditions.

In our time, the olive tree often propagates by means of cuttings or by grafting a one-year-old sown plant. Therefore, in the first case you have a plant with a genuine root, whereas in the second case you may see shoots of the wild olive from the stock once it has been transplanted to the permanent place of growth unless these are cut away regularly. Both types of propaga-

---

18 *Il.* 17.53 ff. describes a young solitary olive tree in the mountains, near a spring. The situation seems not typical, but expressive.
19 Foxhall in Wells, ed., 1992, has suggested that the *hypopremna* are identical with the so-called *ovoli*; see also Hartmann/Kester 1968, 592: 'The characteristic swellings, sometimes called "ovoli" usually found on the trunks of old olive trees may be cut off and planted in early spring. They contain both adventitious root initials and dormant buds; hence new roots and shoot systems can regenerate. This practice is somewhat damaging to the parent tree, however, and is not widely used.'

tion are frequently employed in a nursery from where the trees are transplanted to their permanent habitat. In other places grafting is also used on older self-sown wild trees (Plates 2.8, 2.9). Because by sowing the olive tree reverts to its wild form, a spreading of wild olives takes place continuously as birds eat the fruits. The pits are spread with the droppings from the birds and therefore have a very favourable chance to grow, and when the tree has reached a suitable size of from 3 to 5 feet in height, it is grafted. In many cases, the farmer has stemmed the tree when passing by. Naturally, such wild trees thrive very favourably outside the cultivated area, that is, in the maquis which is used for grazing of sheep and goats. The advantage of this form of propagation is evident. Before being grafted the tree has taken root and is in full growth; therefore, you can avoid watering it during its first year, a work which can be extremely cumbersome in a mountainous terrain far from the nearest supply of water. Actually, planting of trees from nurseries is the easiest way in relatively accessible areas to which water may be transported without unreasonable difficulty, for instance by using donkeys or mules. Presumably, it is also possible to move such self-sown wild trees, with their roots in a clod of earth, in the humid periods of autumn and winter. This is well known from modern Greece; the vegetative method applied in the nurseries lends itself to very safe propagation of trees, the nature and productivity of which is well known. However, the written sources that deal with ancient Greek agriculture do not yield safe clues for us to assume that a systematic mass-production of vine and olive-trees with a view to transplantation took place. Admittedly, the lexicon of Liddell, Scott and Jones does give the meaning 'nursery' as a secondary translation of *phyteuterion*, but if we examine the three contexts which are quoted, the meaning is uncertain, to say the least. In the speech of Demosthenes against Nikostratos (53.15), the latter is accused of having, by night, infringed upon the property (*chorion*) of the plaintiff, of having removed fruit trees, of having cut down *anadendrades* (grape-vine on trees), and of having destroyed the *phyteuteria* of olive trees row by row. It is clear that here *phyteuteria* means, simply, 'young plants', but whether the entire *chorion* is a nursery in the proper sense of the word, intended to furnish plants for later transplantation in a professional way, remains uncertain. Only the fruit trees seem to be in their first stage. The sequel shows that there were also roses in bloom. No building is mentioned. The other two occurrences mentioned are taken from inscriptions, but in one case (*IG* II^2 2493) the word appears in a lacuna so that the meaning cannot be determined. In the second instance (*IG* I^2 94), in a decree it is stipulated that the leaseholder of a sacred piece of land must grow at least 200 *phyteuteria elaon*, preferably more if he so wishes. In this context the word obviously denotes 'plants', but nothing is said as to from where they are to be procured. It is, of course, possible that the leaseholder is expected to take them from whichever supply he may have at his own disposal, or that he is supposed to plant at least 200 cuttings and

Plate 2.9 The same grafting as in Plate 2.8, six months later; the wild olive has been heavily pruned and the domesticated scion is developing

Plate 2.8 Grafting on a wild olive, Methana

then see to it that those that fail to develop are replaced.

It goes without saying that it is precarious to form an argument *e silentio* on the basis of sources so eclectic in their descriptions as those mentioned here, but one cannot help wondering at the lack of explicitness in the writings of Xenophon as well as those of Theophrastus. It should also be mentioned that descriptions of nurseries proper, in Greek, are first found in the Byzantine collection known as *Geoponica* which draws heavily on Roman sources. Here, Cassianus Bassus uses the term *phytorion* to denote the Roman *seminarium*; this term has not been attested previously. We cannot, therefore, exclude the possibility that procuring plants suited for transplantation was an entirely private matter in the Greek world. This should not lead us to assume that there were not large stands of olive trees; everything points in that direction. It is merely a matter of an infinitely less professional division of labour than that we find in the writings of Roman authors when they dealt with agriculture.

Grafting on wild trees makes it possible to include land that would otherwise not be arable. In his novel *Brandy and Roses*, Theodor Kallifatides, a Swedish-Greek author, tells us how in the 1930s a village school-teacher replanted forest on the sides of the mountain and grafted the thousands of wild olive and pear trees that grew around the village Richea between Sparta and Monemvassi. This is a clear case of the inclusion of marginal land into cultivated land without any particular effort, so there is not much doubt that many modern scholars entertain exaggerated notions of the amount of labour connected with a transition to the growth of olives in a country where olives have been grown for centuries. The possibilities of taking a short cut were available to the ancient farmers as well as to those of the present time, but here our sources fail us.

As is well known, once the olive tree has taken root, it can grow to a very old age. Theophrastus says that normally the tree will reach an age of about two hundred years (*HP* 4.13.5), but he knows of specimens which are presumably much older, such as Athene's sacred olive tree on the Acropolis and the sacred wild olive tree in Olympia (*HP* 4.13.2). You can also rejuvenate the tree by cutting it down, allowing the new shoots from the stump to form a new tree. One still sees examples of this, and that it was practised in antiquity appears from Theophrastus (*HP* 2.7.2 ff.) as well as from an inscription *IG* II/III^2 2492, which clearly refers to this practice (Plate 2.10). According to modern accounts describing the cultivation, this makes it even easier to pick the olives.

Whereas Xenophon's description of the olive tree is limited to the planting, Theophrastus is much more informative. This applies particularly to the general passages in *Historia Plantarum* 2 and *De Causis Plantarum* 3.

Pruning is very important (*HP* 2.7.2). In the first place dead branches

Plate 2.10 Old olive tree, cut down, with new shoots growing from the trunk, Crete

must be removed, but with Androtion[20] as his source Theophrastus states that myrtle and olive are the trees that require harder pruning than any other trees except the vine. The importance of pruning for fructification is emphasized. Androtion is also quoted to show that olive and myrtle should be fertilized and watered. Furthermore, the soil should be hoed between the trees, and the roots should be pruned like those of the vine (see especially *CP* 3.8.1). This forces them to grow downwards. Unfortunately, Theophrastus does not tell us how often the olive tree should be pruned. In another context (*CP* 3.7.7) he mentions that the vine should be pruned every year, other trees every other year and others again only every fourth year. Today normal practice is to prune the olive tree very heavily, when necessary, but we are not informed of the practice in ancient Greece. Columella, who is not very interested in the cultivation of olives, recommends to prune only once in eight years (5.9.16). Obviously, pruning of olive trees was not a regular work to be done at a certain time every year.

20 Androtion as an agricultural writer is not well known; see the Loeb edn of Theophrastus, *CP* 1. xx ff. (Einarson 1976).

THE ART OF AGRICULTURE

We have now arrived at a picture of the growing of the olive tree with tilling of the soil and with pruning. The fact that according to Androtion the olive tree thrives with water and fertilizer does not, of course, tell us whether watering and fertilizing was typical. We are also left without recognition of the well-known fact that the olive tree yields substantially only every other year (it is briefly mentioned, *CP* 1.20.3). No precise times for the individual processes are indicated. These have to be deduced from the general descriptions – that is, pruning late in winter before growth begins again, thereafter perhaps fertilizing (*CP* 3.9.1). Since the necessity of continued tilling of the soil is so emphatically emphasized in connection with the description of viticulture, it seems reasonable to assume that crops from other trees are less demanding in this respect.

The picking of the olives today takes place from autumn until early spring. Theophrastus has no description of this, but a well-known vase picture illustrates this part of the work. A man has climbed a tree while others from the ground below poke at the olives with sticks and then collect them and put them in a basket (Plate 2.11). This rather closely resembles modern picking-methods where, however, the fruits may also be brushed off by means of a kind of currycomb. In other contexts Theophrastus uses the word *rabdizein* for taking down fruit from trees by a stick, especially to demonstrate that this might cause some harm (*CP* 1.20.3 and 5.4.2).

Plate 2.11 Picking the olives, Attic black-figure amphora, British Museum

OTHER FRUIT TREES

Apart from the two classical crops of trees, vines and olives, we have mentioned a number of other fruit trees which make their appearance already in Homer's epics. First and foremost, there is the description of the orchard of Alkinoos (*Od.* 7.115 ff.) with pears, pomegranates, apples and figs along with olives. Here we are dealing with formulaic verses, repeated in the description of the torments of Tantalos (11.588 ff.). Cultivation is mentioned in the passage when Odysseus meets Laertes; it is only when the hero is able to recount the number of trees given to him by his father when he was a child that Laertes acknowledges his identity. These are 13 pear trees, 10 apples, 40 figs and 50 rows of vine – in other words an orchard of considerable size laid out far from the built-up area. Alkinoos, on the other hand, has his orchard close to the palace; this is the idealized land of happiness where, like the grapes of the vine, fruits ripen successively (cf. p. 26). There is a grave contrast to this when Laertes, wearing old clothes, appears as fully occupied with his work, probably so as to indicate his fall in social standing.

Xenophon mentions but briefly the planting of the fig tree together with all other fruit trees in connection with the planting of the vine. It is only the olive that requires a deeper plant hole (cf. p. 35).

Theophrastus, of course, also discusses a series of cultivated fruit trees, primarily fig, apple, pomegranate, almond and quince, whereas chestnut, hazel and walnut, for instance, are regarded as growing in their wild form, which does not, of course, preclude that their fruits formed part of the fare. It is a feature common to the cultivated species that their propagation is vegetative; with propagation by means of sowing they revert to more primitive forms, at times even to the wild form. The fruit trees propagate mostly by cutting and grafting. Pruning is necessary. The botanical characteristics are in focus as well as the nature and quality of the xylem, gemmation, blossoming and so on. The specific cross-fertilization of the fig tree by the gallfly, which presupposes the presence of wild fig trees, is mentioned (see especially *CP* 2.9.5). The windfall of unripe fruits constitutes an additional problem which is recognized, and different types of the individual species are referred to. With Theophrastus as our witness we cannot, however, arrive at a definitive impression of the role that fruit-farming played in Greek agriculture; but one piece of information – to wit that certain trees should be planted close to one another, others at a greater distance – would seem to indicate that we are dealing with plantation on a somewhat larger scale.

Jane Renfrew (1973) reviews the prehistoric remains of tree-fruits. Fig, apple and pear were often used as dried fruits, in which case they will stand storing over periods of considerable length. In particular, the dried fig has a very high nutritional value owing to its sugar-content. Naturally, it is a problem how to determine the degree of domestication, especially when we are dealing with apples, pears and various forms of drupe like bird-cherry and plum. In their wild forms, these are found as part of the natural vegetation and were presumably an important supplement to the general fare from the beginning of time. In this context we can be sure that fruit-growing has been known long before the beginning of the historical period, and that like the growing of vine and olive, systematic fruit-growing presupposes a knowledge of vegetative propagation that must be as old as fruit-growing itself.

OTHER CROPS

Apart from grain, many annual plants were grown. We have already noticed that Theophrastus does not include millets with the grains but regards them as part of a separate group of plants that are sown in spring and ripen very rapidly. When he calls them 'summer plants' (*therina*), the reason is presumably that from a botanical point of view the category is very heterogeneous and has no specific botanical designation. Here we are dealing with a group of plants that are also called *anonyma* (*HP* 8.1.1).

Millet (*kegchros, panicum miliaceum*) and Italian millet (*elymos, setaria Italica*) are both known. The species will not suffer the cold of winter but is, on the other hand, very resistant to drought (*HP* 8.7.3) and may reach ripening in forty days (8.2.6). Kernels appear in palaeobotanical contexts, but as a crop it seems that millet was unable to compete with the other types of grain.

Pulse, likewise, is known from palaeobotanical finds in a number of species, such as *kyamos, vicia faba* (broad bean), *pisos, pisum sativum* (pea), *phakos, lens esculenta* (lentil), *erehinthos, cicer arietinum* (chick-pea), *orobos, vicia ervilia* and other species of vetch. In the *Iliad* (13.588), in a simile, there is a description of the cleaning of beans and chick-peas by casting them on the threshing-floor. This would indicate that a considerable amount is involved.

The description of the pulses given by Theophrastus is woven into that of the other annual cultivated plants (*HP* 8). They may be sown in autumn and in spring. The early sowing is usually preferable, but lentil, chick-pea and pea may be sown later like the bean. As in general the pulses have a long period of blossom, they also ripen at different times (8.2.5), and for that reason it has probably been necessary to pick them several times. The individual species are also divided into different sorts described in 8.5.1. Theophrastus is not unmindful of the ability of the pulse fruits to improve the soil (8.9.1). The chick-pea exhausts the soil, whereas the beans are said to improve it. The subject is resumed in *CP* 4.8.1, where the reason is looked for in the brief growing period of the pulses. It is mentioned elsewhere that in Macedonia and in Thessaly you plough in the beans while they are still in bloom for the soil to be improved. It is quite plain that Theophrastus does not know the capacity of the pulses to absorb and store nitrogen from the air, and the use of this form of soil improvement appears to be still on an experimental level, whereas it is fully developed in Roman agriculture. In turn, this raises the question of whether in Greece there was a tendency towards abandoning the two-field system with its alternating fallow and crop in favour of a three-fielded system with rotation of crops, for example, grain – pulse – fallow. The question has frequently been discussed, but there is no unambiguous indication to be found in the writings of Theophrastus – perhaps because, quite simply, this question is not of a botanical nature, but a practical agricultural problem.[21]

Pulse was an important part of the daily diet and undoubtedly played a considerable role, being rich in protein and thereby covering a very large deficiency in a fare where meat was not plentiful. It is a matter of debate, however, whether pulse should be classified as part of gardening rather than agriculture in the proper sense of the word; it must be assumed that over a very modest area many attempted to satisfy their requirements by intensive growing. This applies to a number of vegetables, the growing of which is described in the seventh book of *Historia Plantarum*. These are different sorts of cabbage, beets and turnips, celery, onions and garlic, etc. Furthermore, spices were often collected from wild-growing plants. Recent socio-anthropological research has pointed to the role that collecting wild-growing plants still plays for the Greek peasant population as a natural reserve.[22] Although we do not underrate this supplement to the cultivated produce, we feel that these problems should not be discussed in connection with agriculture.

---

21 Hodkinson 1988, 42, maintains that 'the fact that he [Theophrastus] treats them [the pulses] in the same book as his discussion of cereals and separates them from his examination of garden vegetables and herbs in book 7, suggests that they are viewed as a normal component of field agriculture used in rotation with cereal crops and not just confined to small scale garden horticulture'. We do not think that agricultural view was dominant for Theophrastus and cannot see how one can use this slight evidence to introduce a rotation-system with pulses grown as fodder for cattle, thereby considering Greek agriculture as mixed farming. See further pp. 25 ff. above.
22 Forbes 1976.

3

AGRICULTURAL IMPLEMENTS

The study of the implements of Greek agriculture presents a number of difficulties. No doubt this has contributed to a limited interest in this area. Most implements were manufactured wholly or partly from wood, for which reason few physical remains have been discovered. Whereas bronze is reasonably well preserved, iron is not very durable; for this reason there are cases where we know less about the implements of the historical period than of those that hail from the Bronze Age. In Greek art, pictures of agricultural work are often preserved, but it is obvious that, for example, the Attic vase painters did not see it as their purpose to deliver a workshop drawing of the implements but merely to suggest their presence. Their contemporaries knew of course what was involved. Certain implements that were part of a ritual ceremony are frequently depicted, but often in an emblematic context that does not make it possible for us to see their function. In the literary tradition the difficulty lies in determining precisely which implement is referred to. Frequently the same tool seems to have several names, in the same way that we also assign more than one name to tools for which the Greeks had only one name. It is, however, not very practical to use the same term for the sickle as well as for the assortment of curved knives used for grafting and pruning, and so on. Only the curvature of the blades is a common feature whereas the work carried out by means of these tools is very different. To the contemporary users it was a different matter because naturally they knew precisely which tool was to be used in every specific case.

The agricultural implements can be studied by comparison if you look at modern parallels from non-industrialized agricultures, or a philological and archaeological method can be applied by which the literary evidence is studied in connection with archaeological sources. A combination of these two methods, as far as Roman agrarian history is concerned, will be found in K.D. White, *Agricultural Implements of the Roman World* (1967b), where one author controls the literary and the archaeological material, at the same time having a wide practical knowledge of the use to which such implements were put. A study like this is still a desideratum for the study of Greek agriculture, but in Archaeologia Homerica Vol. II. H, *Die Landwirtschaft im homerischen Zeitalter* (1968), there is a brief chapter by Wolfgang Schiering, 'Die Landwirtschaftliche Geräte', with a thorough discussion of the individual tools mentioned in the poems and a comprehensive set of notes with references to the archaeological material. This study is indispensable for anyone who needs an exhaustive reference to the extensive archaeological literature. M.-C. Amouretti's treatment (1986) is far more ethnographic in its aim. The starting-point is the question, what is functional? There are references to the literary and archaeological sources, but these are not always discussed too thoroughly – perhaps because, in the author's opinion, little new information would be obtained as most implements are mentioned *en passant*. On the other hand, the inclusion of an ethnographic analysis of implements from non-industrialized modern agriculture is often of considerable interest.

In the Index of Greek words, K.D. White lists terms for agricultural implements, but a closer examination will show that most of them are terms from late lexicographic writings. Here, as in an investigation of Byzantine agricultural implements undertaken by Antony Bryer (1986), we are faced with the problem whether the implements have undergone essential changes since the Classical period. In particular, there is a problem as to whether the more advanced agricultural technology of the Romans left its traces in Greece. This can only be determined by means of a comparison between implements that have in fact survived; and as we have mentioned, such specimens are preserved only in a very limited quantity. Thus it becomes a question of an overall interpretation of agricultural history, and a question of the extent to which later material may be introduced.

As far as the recent past is concerned, a series of investigations of traditionally employed implements in Greece is available. We shall confine ourselves to a reference to an excellent catalogue, *Traditional Methods of Cultivation*, from an exhibit in the Benaki Museum in Athens, 1977–8, by Psarraki-Belesióte, where a vast amount of material has been presented. In many cases one should be inspired by modern implements in order to form an impression as to how those of antiquity may have looked, but it would be a jump into the relatively unknown if we were to identify, with any feeling of certainty, ancient terms for implements from specific modern tools. The possibility of local variants is always present. This applies specifically to a number of simple tools which the farmer made from wood in accordance with his own needs and preferences.

Finally, we are faced with the problem of a precise dating of the time when a particular implement was introduced; for we cannot in any way be sure that one and all of the implements known have been in existence from the very moment when agriculture began. Here the difficulty is that an archaeological discovery of, or the reference to, an implement gives us no more than a *terminus ante quem*. We are not likely to arrive at a closer dating of what may have been a technological innovation within the limits of the tradition of our sources.

It is not our intention, in the present study, to discuss all the aforementioned aspects in each individual case. It must suffice to account for the most important implements and their functions. It would seem reasonable to group the individual implements that have to do with improvement of the soil and preparation of the soil for sowing. We shall then follow the agricultural year and the individual tasks as we did in Chapter 2.

THE PLOUGH

The ancient plough has often been discussed and interpreted. We owe this to the fortunate circumstance that we have a reasonably accurate literary description in Hesiod's *Works and Days*, ll. 427 ff., together with numerous depictions of ploughs and ploughing, on statuettes in terracotta or bronze, on vase paintings (especially Attic black-figure vases) and finally on coins and other miniatures.[23] There can be no doubt that the plough was symmetrical, that is to say, it could not turn the soil in the modern way; instead, it left a scratch in the soil. In other words it was an *ard*, and as such it has been in use in the Mediterranean area until our time, when it may still be found, mostly in mountainous areas. The function of the *ard* is partly to destroy the weeds, partly to air the top-soil so that it will become sufficiently porous for the plants to take root. This double function has been admirably described by H.A. Forbes (1976a), who also emphasizes the effect of moisture retention following the tilling of the soil.

As described by Hesiod, the *ard* consists of the following parts: the beam (*gye*) is a curved piece of wood connecting the sole (*elyma*) with a drawbar (*histoboe*) on to which the draught-animals are hitched with a yoke (*zygon*). A stilt, equipped with a handlebar (*echetle*), may be attached to the sole. In front, the sole may be equipped with a ploughshare made of bronze or iron (*hynis*) (Figure 3.1). The ploughshare is not mentioned in the poem, but it is apparently known by Homer (*Il.* 23.834 ff.). Archaeologically it is attested in bronze as well as iron. Hesiod recommends that you should have two ploughs, the composite version and the non-composite version, the *autogyon aratron*. The latter may be difficult to interpret, but it is possible that sole and beam consist of one piece of curved wood to which the drawbar is attached directly. Amouretti is undoubtedly right in assuming that we are not dealing with two types of ploughs but with variants of the same implement, depending on the material available: Hesiod has his farmer go to the forest in order to find suitable wood. As draught-animals Hesiod

---

23 Drachmann 1938, with references; White 1967b. Kothe 1975 demonstrates how differently the details have been interpreted. Amouretti 1986 gives a full discussion with many illustrations and much ethnographical material.

AGRICULTURAL IMPLEMENTS

recommends 9-year-old oxen, and the ploughman should be an experienced man of 40 years of age, someone who does not look towards his fellow ploughmen but is able to plough in a straight line.

In the great epic ploughing is frequently used in similes. The fact that the work is strenuous to the oxen as well as to the ploughman is mentioned (*Il.* 13.702 ff. and *Od.* 13.31); and the challenge offered by Odysseus to Eurymachos (*Od.* 18.371 ff.) shows that steering a straight course for the furrow is equally difficult. In one case, ploughing with a team of mules is mentioned; these move faster (*Il.* 10.351). The most famous portrayal of ploughing is the description of the shield of Achilles (*Il.* 18.541 ff.).

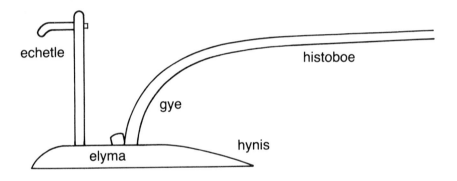

Figure 3.1 Diagram of a plough (after Drachmann 1938)

As we have mentioned, depictions of the plough occur frequently. We show a well-known terracotta statuette from Boeotia, now in the Louvre (Plate 3.1). Here all the parts of the plough mentioned by Hesiod are to be found. It may without reservation be compared with a modern *ard* photographed on Mykonos in 1978 where it was still in use (Plate 3.2). From among vase paintings we reproduce a black-figure kylix showing a series of agricultural activities, among them two scenes illustrating ploughing, one employing oxen, one mules (Plate 3.3). It may be noted that here, as frequently, ploughing is associated with sowing. We choose to interpret this as an artistic convention that emphasizes that autumn ploughing and sowing

Plate 3.1 Plough, Boeotian terracotta, Louvre

Plate 3.2 Modern plough (*ard*), Mykonos

are closely associated in time. It may further be noted that here both ploughmen seem to stand with one leg on the sole. Alternatively, it is possible that they are in fact walking beside the *ard* – that is, behind it, as seen from the point of view of the spectator. This lends a considerable movement to the representation. In the well-known vase-painting of ploughing by Nikosthenes, the ploughman stands behind the *ard* holding a long stick with which he steers the oxen.[24]

According to Hesiod there are three ploughings, one in spring, one in summer and finally one in autumn immediately preceding sowing. This, of course, presupposes that we are concerned with a fallow-field system. As we have seen, the biennial system seems to have been the usual procedure. There is, in fact, a verb (*neao*) which is used specifically for the ploughing of a fallow field.

HOE AND SPADE

Apart from the plough which employs animal traction, manpower in itself was also used in the tilling of the soil. *Neain* and *skaptein*, for instance, occur in Theophrastus *De Causis Plantarum* 3.20.1, but in spite of this we do not know with any certainty whether the Greeks used a spade when 'digging' – usually we translate *skaptein* and sometimes *oryttein* by this verb. It is the weight of the digging person when he places one foot on the spade that forces it into the ground. Most ancient spades are from the Roman period and have been found in the northern provinces. The Mediterranean soil, which often tends to be stony, is more suited to be prepared with a hoe. The fact is that we cannot with certainty find any Greek word as a term for the spade, whereas it does occur in Latin. This may be fortuitous, but it is a circumstance that merits some afterthought. Nor are there any safely identifiable remains of spades from the Greek area. However, this may be just as much a coincidence as the fact that the word does not occur in any clear context. It is possible that a wooden spade, perhaps with its blade reinforced by an iron shoe,[25] may have been in use, but there is no evidence to support this assumption.

As for the hoe, the case is different. It is most often known as *makele*, *makella* and *dikella*, the latter – according to the name – the two-pronged hoe. Both tools are used for a thorough tillage of the soil, and Theophrastus maintains that the *dikella* is better than the *ard* in breaking up the weeds from the fallow field (*CP* 3.20.8). It is probably an implement like this that Xenophon has in mind when describing the treatment of the fallow (*Oeconomicus* 16.15), and not a spade, although the expression *ei skaptontes ten neon poioien* is applied. In a simile Homer describes a man leading water

---

24 Berlin, inv. no. 1806 (Beazley 1956, 223), details given by Amouretti 1986, 84, fig. 10.
25 Amouretti 1986, 93, stresses the frequent use of the hoe in the eastern Mediterranean region in modern time; for the Latin *pala* see White 1967b, 17 ff.

Plate 3.3 (i and ii) Agricultural labours, Attic black-figure kylix, Louvre

Plate 3.3 (iii and iv) Agricultural labours, Attic black-figure kylix, Louvre

through a water conduit to trees and a garden with a makella in his hand (*Il.* 21.257 ff.); it is probably a broad-bladed hoe that is meant here, but the hoe shown on the Louvre vase has the appearance of being much more pointed (see Plate 3.3). These two implements could, of course, have been known under the same name. The two-pronged hoe is known from few examples in Archaic or Classical art. In vase paintings showing scenes from a palaistra we often observe the pickaxe, and Schiering is undoubtedly right in assuming that it may have been used to smooth out the field if the side of it is used for levelling.[26]

In addition, there is the term *sminye* which also denotes a hoe but it is difficult to determine the possible difference between these two terms, or whether they are perhaps synonyms denoting one and the same tool. It is certain that hoes must have existed in many variants and of different weights in order to serve various purposes. When, for instance, Hesiod prescribes for a boy to hoe down the seed with a *makele* immediately after the sowing (*Works* 470), there is obviously no reason to assume that a heavy hoe, suitable to break up the fallow, is involved. This would be a waste of effort. Likewise, weeding and hoeing of the cornfields must have been undertaken with a very light hoe, perhaps of the kind that Xenophon calls a *skalis* (*Oec.* 17.15), also found elsewhere in our literary sources. The same tool was probably used for that weeding of the grain recommended by Theophrastus, *skalsis* (see above, p. 25), but exactly what it looked like we do not know.

SOWING IMPLEMENTS

Vase paintings show clearly that this part of the work required merely a sack or a basket with a strap round the neck so that scattering was done by hand (see Plate 3.3). As far as we know, a harrow was not used, but the seed could always be put down by means of a small hoe, as mentioned above.

SICKLE/CURVED KNIFE

The standard tool for harvesting was the sickle (*drepanon, drepane* or *arpe* as Hesiod calls it, *Works* l. 473). In Homer, it is also used for haymaking (*Od.* 18.366 ff.), there with the epitheton *eukampes* (that is, 'well-' or 'beautifully bent'). The large hay-scythe is an invention that belongs to a later period in antiquity; presumably by then fodder was required in much greater quantity than in earlier times for horses of the heavy cavalry.[27] The most authentic portrayal of the harvesting of grain is found in the description of the shield of Achilles (*Il.* 18.550 ff.) where not only the mowers but also the binders of the sheaves are mentioned. The large-scale farming suggested in this context

---

26 The *dikella* is seen on an amphora stamp from Thasos, Amouretti 1986, Plate 13; for the use of the pickaxe see Schiering 1968, 153, note 1148.
27 White 1971–2.

AGRICULTURAL IMPLEMENTS

may presumably be attributed to the general drift of the epic. As far as the mowing *per se* is concerned, it bears comparison with Xenophon's description in *Oec.* 18.1 ff., where the technique is clearly accounted for with no mention of the sickle at all (cf. above, p. 25).

Sickles are found frequently in archaeological contexts; as, for example, in a considerable collection in Perachora where Dunbabin (1940) feels that, like spits (*obeloi*), they may have been used as media of exchange in a pre-monetary economy. There is a similar hoard from the Rheneia tombs, and some from Corinth and Olynthos[28] (Plate 3.4). They have often been published designated as sickles and sometimes catalogued as pruning- or gardeners' knives. From the fragments preserved it is often difficult to determine whether it is one implement or the other which deserves the identical term, but it is not easy to trace a well-preserved specimen of the grain sickle. The curved knife is known primarily from a series of Spartan victory-inscriptions from the sanctuary of Artemis Orthia where a bill hook has been fixed on top.[29] Here, in any case, one is not left in doubt with regard to its purpose and function in agriculture, nor were the contemporaries when they were about to use the tools in practice. What was meant by the representation remains uncertain. With regard to the harvest sickle it cannot be determined whether it was usually serrated as it is known from elsewhere; there are but few indications which point in that direction.

THRESHING-FLOOR

The threshing floor (*aloe*) is mentioned frequently and has been so ever since Homer, but we are told very little about its structure or form. Here we are forced to consider analogies with well-known modern threshing-floors which, nowadays, are seldom used (Plate 3.5). They are circular and usually have a hole in the centre; into this is fixed a pole which can be turned. The draught-animals are hitched to the pole so that they keep the right distance; this, of course, can be varied. The threshing-sledge is not mentioned in the Classical period and is perhaps a later invention.[30] The kernels were trodden out by the draught-animals.

Occasionally, ancient threshing-floors have been discovered *in situ*. As an example let us mention that near the so-called 'Princess Tower' at Sounion in Attica, archaeologists have observed

> a terraced and carefully paved circular platform nearly twenty meters in diameter, encircled at the east by a low rim of stones, at the west by a careful cutting in the native rock. . . . The circular platform is certainly an ancient threshing-floor.[31]

28 References in Schiering, 1968, 156, Amouretti 1986, 100.
29 Dawkins 1929, 229–39, fig. 133; cf. Steinhauer, n.d., 18.
30 White 1967b, 152 ff. Foxhall has suggested that it could be identified with a word in the Attic stelai, *okistion*; see further Lohmann in Wells, ed., 1992.

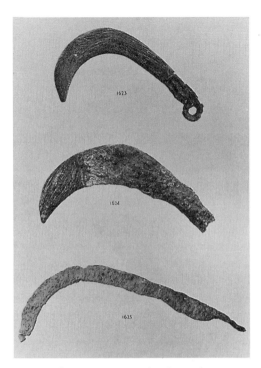

Plate 3.4 Pruning-hooks and reaping sickle, Olynthos

Plate 3.5 Threshing-floor, Methana

A corresponding structure has been found not far from there near the 'Cliff Tower' and in other contexts as well. Although we cannot be quite certain about the dating of them, it does seem that everything points in the direction of ancient relics.

WINNOWING SHOVEL AND WINNOWING BASKET

Liknon is the winnowing basket by means of which the grain is winnowed when grain is thrown from the basket upwards and caught when it falls down. When larger quantities are involved, the grain is thrown against the wind with a shovel (*ptyon*) so that the chaff will be blown away. This is described in a couple of Homeric similes (*Il.* 5.499 ff. and 13.588 ff.), the latter passage dealing with the threshing of pulse. The winnowing basket, as well as the threshing shovel, are discussed in a famous article by Jane Harrison (1903, 1904) dealing with the entire process of threshing as being closely connected with the cult of Demeter.

OTHER IMPLEMENTS

A number of other tools may be added to this somewhat limited list. Hesiod mentions the mortar with its pestle, *olmos* and *hyperon*, both made from wood, and a waggon mostly manufactured by the farmer himself (*Works*, 423 ff.). The mortar is shown on vase paintings, but this is not the place to enter into a discussion concerning the Greek waggon.[32] It is enough to note that much transport took place on the backs of mules or donkeys, a circumstance that has the distinct advantage of not depending on a carefully laid out system of roads. In addition, there are tools made from wickerwork and clay, baskets designed to collect grapes and olives and large storage vessels for the preservation of grain and olives. From the time of the Geometric style we possess a number of intricate vases that have been interpreted as models of storage vessels for grain, a type of silo. These have been well described in archaeological literature,[33] but it may be difficult to envisage exactly how they corresponded to functions in actual peasant life. Naturally, the farmer also had for domestic use saw, hammer, axe and other hand-tools, some of which served him for the felling of trees; these could also be used for the pruning of fruit trees. If we are right in assuming that grafting was widespread, this industry would also require special tools, particularly very sharp grafting knives and corresponding whetstones. All these tools, however interesting they may be, we shall disregard in this connection because in many ways they may be said to be of marginal interest

---

31 Young 1956b, 124; Lohmann in Wells, ed., 1992.
32 Richardson/Piggott 1982.
33 Coldstream 1977 s.v. 'granaries'.

for the understanding of agriculture as such. It remains for us to accept that Greek agriculture seems to have been relatively poor with regard to implements. In a poem from the *Anthology* (6.104), Lysixenos hangs his tools in a sacred place, to wit, the seed bag, the hammer (*sphyra*) by means of which lumps of earth are broken, the curved harvesting sickles, the plough, the *histoboe* as well as the plough itself and the ploughshare, the stilt and the three-horned wooden pitchfork. The only implement which points towards a post-Classical era is the threshing sledge, *tribolos*. With this we may compare the very brief inventory which we find among the poor peasants who have moved to an outlying place in Euboea (Dio Chrysostom, 7.42).

IMPLEMENTS FOR MAKING WINE AND OIL

Whereas the tilling of the soil in vine and olive plantations has been performed with implements identical with those used in the grain fields (that is, particularly the hoe and sometimes the *ard*), the harvesting of vine and olive, together with the pressing, naturally requires specific tools. As far as the produce of wine is concerned, we possess ample evidence, mostly archaeological material which has been well studied and examined. In his commentary on the Attic stelai, Amyx (1958) discusses a number of these implements, and B.A. Sparkes (1976) has collected and discussed a very large number of vase representations with subjects from vintage and wine pressing. As a rule satyrs, often crudely depicted, are treading the grapes. Perhaps this makes the description less realistic, but it does allow us to make deductions as to how things were executed in real life. Harvesting of olives and pressing has been much more neglected in art and, as we shall see, a number of unsolved problems remain.

Vintage is mentioned by Homer (*Od.* 7.123–5) and by Hesiod (*Works*, ll. 612–14). It has been a matter of some speculation that in the latter description the grapes were dried for several days, but presumably the reason is that they should be conserved so as to turn into raisins rather than be left to dry prior to pressing so as to produce wine. The reference from the *Odyssey* also mentions treading grapes; and the description of the shield of Achilles in the *Iliad* refers to the collection of grapes in large wickerwork baskets. This is exactly what the vase painters describe. In the following, we shall refer to the Greek terminology as compiled by Sparkes, although it cannot at all times be completely taken for granted. For instance, where the identification depends on an explanation offered by a late lexicographer, we cannot be sure that the explanation is valid for that much earlier period which we endeavour to describe here. Inasmuch as the terminology has found general acceptance, and seems reasonable, it will thus be rendered here.

The grapes were picked and collected in large wickerwork baskets, perhaps *phormoi* or *kophinoi*, which were then carried to a wooden pressing-board (*lenos*) (Plate 3.6). The latter often had a spout and was placed on four supporting legs, perhaps sloping slightly forwards. The grapes were placed in an open basket or sometimes in a closed wickerwork sack, occasionally with a handle added to it, and this was worked alternately with feet and knees, so that the juice seeped on to the board and flowed downwards into a container (*hypolenion*). In a number of cases we also see a person treading in a large vat, and can assume that here, too, we are dealing with wine-making, although there are no traces of grapes. Sparkes suggests that when the grapes are trodden in a basket the must is separated from the grape-skin and pips, and ferments separately, whereby a white wine is obtained, whereas treading in a vat might suggest that must is not separated from the grape-skin, in which case you obtain a red wine. This is a possibility. It may be noted that often the treading seems to take place out of doors, close to the vineyard and the picking itself. This, of course, may be an artistic convention linking the two processes together physically, but the pressing-board can be moved without much effort, and one may well imagine that the pressing takes place in the close vicinity of the vineyard so that any waste that might occur during a lengthy transport of the grapes is avoided. In other cases the person who performs the treading holds on to a strap above, which would indicate that the pressing takes place indoors with the strap hanging down from the ceiling. This would apply in cases where the treading takes place in a large vat, as an earthen vat of such dimensions could not easily be moved. A couple of red-figure kraters show a combination of treading grapes in a vat as well as on a pressing-board (Plate 3.7). Here the straps indicate that both processes take place indoors. It is not until the end of the fifth century that a pressing-board made of stone is shown on an Attic red-figure krater, now in Bologna (Plate 3.8). This goes well with the fact that, in the Attic stelai, the term *lenos lithinos* is attested (Plate 3.9). Treading on a hard surface is an obvious advantage, and the wooden pressing-board has the appearance of being rather fragile. Furthermore, it is important to emphasize that pressing, as long as wooden pressing-beds were put to use, would not necessarily leave any archaeological trace to testify that wine-pressing has taken place. No particular installation is necessary for this part of the job, except perhaps a vat dug into the ground in order to collect the must. This ferments in large storage vessels called *pithoi*, and once the fermentation process has been terminated, it is transferred to the commonly used wine jars, the *amphorai*. We know little about this part of the process, but it may be assumed that it was associated with the first day of the Anthesteria festival in Athens, the so-called *pithoigia*, when for the first time the new wine was tasted (see Figure 11.1).[34]

The picking of olives is well known, particularly from the depiction on an Attic black-figure amphora which we have mentioned above (Plate 2.11). It shows two men with long poles, beating olives down from the tree, while a third has climbed the tree and is pushing the olives down; a fourth man is

---

34 Deubner 1932, 93 ff. For an earlier opening of the *pithoi* see Chandor 1976, 119.

Plate 3.6 Picking and treading the grapes, Attic black-figure kylix, Cabinet des Medailles, Paris

Plate 3.7 Treading the grapes, Attic red-figure krater, Museo Civico, Ferrara

Plate 3.8 Treading the grapes on a stone pressing-bed, Attic red-figure amphora, Museo Civico, Bologna

Plate 3.9 Stone pressing-bed, Olynthos

busy collecting them from the ground, putting them into a basket with handles. The poles, or sticks, may well have been reeds, as they are still used for this purpose in modern Greece.

The olive fruits have to be crushed before they can be pressed. For this purpose, an olive crusher was used (*trapetum* in Latin) (Figure 3.2). The difficulty lies in dating this invention. Although the term is obviously of Greek origin, no reference to an oil-mill is extant in Greek literature. It

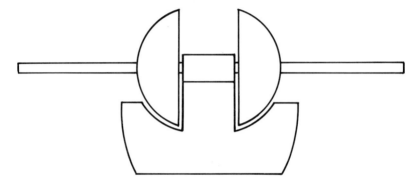

Figure 3.2 Diagram of olive-mill (*trapetum*) (after Drachmann 1932)

consists of a large saucer (*mortarium*) with a cylinder (*miliarium*) in its midst; to this, two millstones (*orbes*) could be fastened. The technical details are communicated by the Elder Cato.[35] Somewhat later, in Imperial Rome, a *mola olearia* is also brought into use; its *orbes* are different, and they recur in the modern olive-mills with one, two or even three millstones. *Mortaria* are not rare in Greece, often standing free in the landscape. Occasionally, the *miliarium* has been cut away in order for the saucer to be used as a water trough. These, of course, cannot be dated. Now and then *orbes* are found in the course of excavations. The earliest specimen is from Chios, found in a context that dates it to the end of the fifth century.[36] There are five specimens in Olynthos, but not found *in situ*. Since this town was destroyed in the year 348, we are furnished with an excellent *terminus ante quem*, and as it had been resettled by new colonists in 433, it is fair to assume that this provides a reliable *terminus post quem*. While the excavators found no *mortarium* in the town, there is now a well-preserved *trapetum* within the excavated area (Plate 3.10). If it is authentic,[37] this is where we find the

35 Drachmann 1932 is only concerned with the more sophisticated machines, but White 1975 has an appendix on the making of olive-oil, where he discusses crushing before the invention of the *trapetum*. See further Forbes/Foxhall 1978.
36 Boardman 1958–9.
37 Runnels has suggested in a letter that he does not believe that the *trapetum* mentioned originates from the excavations of Olynthos, but rather from a Roman villa outside, but Foxhall, who has studied the photographs, believes that it could as well be from the time of the city.

earliest fully preserved *trapetum*, and we may well conclude that the machine dates from the latter half of the fifth century, contemporary with the millstone from Chios.

Now the question arises, how were olives crushed before the invention of the *trapetum*? Some have suggested that they may have been crushed by people wearing very heavy shoes. The derivation of the word *trapetum* might point in that direction, since it is associated with the verb *trapein* which, by Homer and Hesiod, is used for treading the grapes (*Od.* 7.125; Hesiod, *Shield* 301). One might also imagine that they were crushed in a mortar, but this would undoubtedly have involved difficulties. A third possibility is that the olives were placed on a firm underlying layer, whereupon a cylinder was rolled over them. This seems to have been the case in a house excavated in Praisos on Crete, datable to the Hellenistic period.[38] As so often we have to accept that a technological innovation like the *trapetum* is not immediately widely adopted nor does it oust other forms of technology. As is the case in under-developed countries today, earlier methods will have been adhered to slightly obstinately, because the *trapetum* and its replacement, the *mola olearia*, required ready money and a good deal of initiative. It is manufactured from volcanic material, and examination has shown that Aegina and the Methana peninsula supplied materials for many of the grain- as well as for the olive-mills, but exactly when this machine was developed we cannot determine, nor how widely it was put to use before the Roman period, when it was used extensively. We cannot even be sure whether, at all times, it would have two *orbes*, or whether it would have only one.[39] The Chios discovery cannot be decisive on this point because the area was re-populated in Roman times, and a millstone or two may well have been lost. Therefore, we shall have to leave open the question of how the crushing of olives was performed as far as the Archaic and the early Classical periods are concerned. This is regrettable, for it is clear that at the very time when large-scale production of olive-oil intended for the market existed, the crushing of the fruits may well have presented a bottle-neck that could easily delay the entire procedure.

After the crushing, the olive fruits are to be exposed to very heavy pressure. This may be done in a variety of ways. The simplest method consists of placing the crushed olives in a woven bag or a basket, stacking one layer on top of the other and placing a heavy object on top (a stone, for example). Such simple methods are known from modern times, but they are not referred to in our sources, which are regrettably silent on such essential details of the process. The only representation of an olive-press is to be found in a black-figure skyphos in the Museum of Fine Arts in Boston

---

38 Bosanquet 1902, 231. The plan of the house is often reproduced, see e.g. Skydsgaard 1987.
39 Forbes/Foxhall 1978 put the *orbes* from Pindakas and Olynthos together. They are inclined to accept that the *trapetum* from the beginning had only one *orbis*, but Runnels 1981 is not convinced.

Plate 3.10 Oil-mill (*trapetum*), Olynthos

Plate 3.11 Oil-mill (*trapetum*), Pompeii

(Plates 3.12, 3.13), to which we may add a similar (unpublished) representation in the museum at Thebes. What we observe here is a press-bed made of wood, much like the one we know from the pressing of wine; on top of the bed a number of bags are stacked. The pressure is applied by means of a pressing-beam which, assisted by heavy counterweights, is forced down on to the stack. A man stands at the end of the beam while another jumps up on to it, forcing it down by his weight. It is not indicated how the beam was fastened at the end. The motif is shown on the obverse as well as on the reverse of the vase with slight variations.

Here we are faced with a type of press in its simplest form. The representation gives no hint as to where the press is set up, but the pressing-beam must by necessity have been fastened to something at the back. The press is made of wood, so it would leave no archaeological traces. Later, after the transition to pressing-beds made of stone, naturally such specimens can be found; this is the case in the house in Praisos where a stone pressing-bed could be placed near a wall with a hole in which the pressing-beam was fastened. It would seem reasonable to assume that the wooden press was fastened in the same way. In that case the pressing must have taken place indoors, not necessarily in a house or a farm building, but in any case in some sort of shed or shack with a wall sufficiently solid to serve as a counterweight against the considerable upward pressure that the pressing-beam brought to bear at the other end of the beam.

A press of this type can be used for pressing olives as well as grapes, and an important element would be the bags in which the fruits are placed during pressing. A Hellenistic relief shows a stack of such bags under a press. In spite of the fragmentary state of the relief, it is quite clear what is going on (Plate 3.14). Bags of different types and made from different materials are known to this very day wherever olive-oil is pressed, made from wickerwork, jute or even plastic. The decisive factor is that they should be permeable for the liquid pressed, but able to retain solid material, including the stones. The liquid pressed consists of oil and water. If left for a while, the oil will collect on the surface and may be skimmed off unless a particular type of vessel is used whereby the water may be drained off from the bottom. Clay vessels of this type are known from the Bronze Age and later and provide definitive proof that olives were pressed.[40]

The pressure to be applied by the pressing-beam can naturally be achieved in a variety of ways – on the Boston vase, a counterweight is shown. It seems that the weight consists of sacks in which stones could be placed. Often stones are found as counterweights, the stones being tied to the bar with a rope. From Cato's description of the press we find that he still used a drum with handspikes by which the pressing-beam was pulled downwards. In the Hellenistic period, the screw-press was invented, which occupied much less

---

40 Forbes/Foxhall 1978, figs 17–18.

Plate 3.12 Olive-press, Attic black-figure skyphos, Boston (M.H. Hansen)

Plate 3.13 Olive-press, Attic black-figure skyphos, Boston (M.H. Hansen)

Plate 3.14 Bags with crushed olives under the press, relief, British Museum

space and could be combined with the pressing-beam press so that this could be screwed down. But all these variations belong to the Hellenistic and Roman periods and are, therefore, of lesser interest in our context.[41] The development of various types of presses that have been in use up to our own time shows, however, that technical problems have been tackled in different ways. Two completely different types of presses have been combined, to wit, the screw press and the press provided with a pressing-beam; thereby a model was arrived at that was adequate until the industrial revolution, when, by means of hydraulic presses, a much greater (and presumably a more even) pressing was obtained. Still, there is no reason to believe that the press of the Boston vase was not reasonably effective, and it is the only type of press which, with absolute certainty, we can trace back to the Archaic and Classical time.

It is not certain how the task of pressing was organized. Such Roman villas as have been excavated indicate that presses were installed at each farm, but this is by no means a necessity. In a famous anecdote about Thales, the philosopher, we are told that he foresaw a rich harvest of olives and therefore, well-ahead of time, he hired as many olive workmen as he could

41 Skydsgaard 1988a.

get hold of when they were not in demand so that, in the time to follow, he virtually held a monopoly (Aristotle, *Politics* 1259a).[42] Whether they were casual labourers or people who also owned an olive-press, we cannot know, but it would be reasonable to assume that their work was centred round certain places with the necessary equipment. A couple of examples where presses have been discovered by excavation might indicate that these were often located in relatively urban districts as, for example, the Praisos press-room.[43]

With regard to tools, or implements, therefore, we may conclude that proper hand-tools, including the plough, seem quite static, with few identifiable variants, whereas presses and olive-mills seem to undergo a series of more essential changes during the late Classical and Hellenistic periods. This may of course be due to our insufficient evidence, but this explanation is scarcely adequate by itself. The very production of wine and oil is a case of an industry developing from agricultural products, frequently with a view to marketing, and the multitude of containers for wine and olive-oil goes to show that they were often transported far afield throughout the Mediterranean area. For that reason, it undoubtedly paid to improve working procedures here, whereas simpler manual tools rapidly attained a functional excellence that did not prompt any attempt at improving them in any fundamental way. Were you to visit a Greek market today, you would find a number of tools the shaping of which has not changed noticeably since antiquity. This is not intended to suggest that we are dealing with a retarded material culture: rather, to show that at a very early time man developed a technology adjusted to his environment.

---

42 The MSS give two possible variations, and, therefore, we cannot see if Thales rented the workshops or hired the workers. If the first reading were correct, it would be more illuminating in this context.

43 Jones/Graham/Sackett 1973 have an excellent survey of 'items of farm equipment' (see p. 418, note 141). A pressing-room on Delos was published by Bruneau/Fraisse 1981; see further Amouretti 1986, who dates the *trapetum* to the Hellenistic period, see esp. p. 165.

4

AGRICULTURAL BUILDING

How did the ancient Greek farmers live – what were the physical surroundings within which their production took place? The simple question is, in fact, quite difficult to answer. If we turn to our literary sources, they prove to be of little value. We shall refrain from entering into a discussion of the Homeric palace. No one has so far succeeded in obtaining a reasonably accurate harmony between the descriptions offered by the epic and the material remains.[44] This is scarcely odd since the great epic emphasizes the grandiose surroundings with which the heroes were faced. Nor are the descriptions given by Hesiod very accurate. He seems to be at home in some sort of village with a smithy and a hall (*lesche*, cf. *Works and Days* ll. 430 and 493), but offers no descriptions of a farm; Xenophon offers even less. As Ischomachos walks through his house together with his young bride, he does point out storage rooms for grain and wine (*Oeconomicus* 9.3), but he says nothing about rooms for presses or stables. It seems that we are dealing with a townhouse, where the farm buildings are located elsewhere. Here and there in our literary sources there are references to farms, but we never find a precise description. Presumably, it is taken for granted that the reader, or the listener, understands the references without elaboration. When we compare the careful discussions of the most appropriate placing of farm buildings as handed down by Roman authors in their writings about agricultural buildings, the silence in Greek tradition in this matter becomes noteworthy.

Agricultural constructions have often been discussed in recent literature. Jan Pečírka discusses the outlying farms in his chapter, 'Homestead farms in Classical and Hellenistic Hellas',[45] in which he presents an excellent survey of the literary as well as the archaeological material. This study remains an important basis for further research. As for Attica, J.E. Jones (1975) has put together a very useful comparison of private houses in town and in the countryside, and the Attic material is again reviewed by Robin Osborne (1985a); in subsequent studies he extended the geographical horizon.

Before a detailed discussion of buildings excavated, it is necessary to

---

44 Wace/Stubbings 1963, 489.
45 In Finley 1973b, 113 ff.

outline some preliminary conditions. By itself, it is difficult to determine whether a particular building is necessarily associated with agriculture when we consider that most implements were wooden tools. It is only when the stone pressing-bed is introduced that we have solid installations, just as the *trapetum* gives us a fixed point for an interpretation. In our opinion both innovations belong to a time not earlier than the end of the fifth century, and we cannot expect these inventions to have spread rapidly. As we have mentioned, the *trapetum* is often found freely in the landscape with no remains of buildings. It may have been covered by a temporary roof or a hut, but this would not have left traces of any consequence. Investigations of olive- and wine-presses with solid installations all date from the Hellenistic period, and they are often found in more or less urban areas. Dörpfeldt (1895) excavated press-rooms on the western slope of the Acropolis, and a handsome specimen from a town has been found in Delos;[46] likewise, a house with a press in Halieis in Argolis cannot be described as a farm.[47] We have already mentioned the townhouse in Praisos which, similarly, is not a farmstead. All of these are Hellenistic. In fact, we cannot expect to be able to identify agricultural buildings on the basis of their solid installations, nor are storage vessels – for instance the large half-buried *pithoi* – a reliable indication. It stands to reason that the crops need not be stored in the place where they were produced.

Next, one might imagine that the placing of the building would be significant. Isolated rural buildings might suggest that agriculture was connected as an industry. As we have seen, some buildings have a threshing-floor in the vicinity, and if it can be demonstrated that buildings and threshing-floor are contemporaneous, this would be a convincing argument. We have mentioned the two farmstead conglomerations near Sounion associated with a threshing-floor (pp. 53–5). Here the buildings consist of a house, a yard and a tower. This raises the question of whether other towers are associated with agricultural productivity. Young (1956b) is inclined to this assumption and interprets the Greek word for tower (*pyrgos*), as it occurs in the literary sources, as the term for an agricultural holding. He writes as follows, 'We are, I believe, safe in assuming that the Greek country estate comprised three basic structural elements, tower, court and house.'[48] Osborne and others, however, argue that the tower may serve widely different purposes, sometimes, although not always, affiliated with agricultural activities as such.[49] Therefore, we are no longer at liberty to conclude that any connection between architectural typology and agricultural function exists.

Outlying farmyards, near the cultivated fields, are of course merely a

46 Bruneau/Fraisse 1981.
47 Boyd/Rudolph 1978.
48 Young 1956b.
49 Osborne 1987, 63.

possibility. Quite possibly, farmers – like so often today – would have their residences in towns or villages. We have observed press-houses in Athens as well as in the city of Delos, but there is little doubt that to a very large extent the population of Athens, for instance, lived outside the walls of the city. The brief remarks of Thucydides concerning the evacuation of Attica at the outbreak of the Peloponnesian War support this view (2.14): here, it is stated, the difficulty with evacuation was that most people had always lived in the countryside (*en tois agrois*). Unfortunately the expression is couched in very general terms, and we cannot tell whether it refers to widely spread homestead farms or rather to the demes (in which case we are dealing with a group of villages). In all events, it is clear that *synoikismos* was political rather than physical. This is not the place for us to review the deme problem again, particularly as this topic has been dealt with so thoroughly by Osborne. It is a matter of some wonder that so few physical remains of the demes have been excavated, but in many cases later occupation still covers the sites, and with the rapid expansion of Athens in the present century much archaeological material from the surrounding area must inevitably have been destroyed. It cannot be denied, however, that archaeological interest in graves and sanctuaries may have played its part in the situation, which still remains to be clarified.[50]

A different possibility with regard to building development would be that larger tracts were made arable by a parcelling-out of new fields, possibly with the farm buildings adjacent to the developed land. Such examples are well documented at Metapontum in Southern Italy and from the Crimea. In the first case, aerial photographs and subsequent excavation revealed a network of fields, and a number of outlying buildings must be interpreted as agricultural properties. Excavation of individual buildings testifies to a development in this fashion from the sixth century.[51] In the Crimea, large mounds of ruins have lain waste since antiquity; only in more recent periods have systematic investigations been carried out. Here we find plots of land of considerable size (*kleroi*) surrounded by walls interlinked by a rectangular system of roads, and each *kleros* is provided with a house with accommodation for habitation, storage rooms, sometimes rooms for wine-pressing, and so on. Some of the buildings have a proper tower. Russian investigations, furthermore, demonstrate that grain, vine and various fruit-trees were grown. The individual plots of land are sizeable, up to 30 ha. per lot. The entire system can be dated to the end of the fourth century and later; the site was abandoned towards the end of the second century.[52]

It now seems clear that, in these two cases, we are faced with agrarian structures that had been laid out very deliberately. As far as Metapontum is

---

50 Lohmann in Wells, ed., 1992.
51 Adamesteanu 1973; 1974, 66 ff. Further excavations by the University of Texas are to be published.
52 Pečírka 1970, Dufková/Pečírka 1970.

concerned, this may be connected with the expansion of the territory of the Greek colony from the sixth century, whereas the extension and fortification of arable soil in the Crimea took place a couple of centuries later. In both cases it would seem that new land has been laid under plough, or a drastic reorganization has taken place – as could be expected when a Greek town is founded on foreign soil where no consideration for former inhabitants was called for. This is the situation described by Homer (*Od*. 6.4 ff.) where the hero Nausithoos founds a city with a wall and distributes the fields to the people. Conditions were, of course, different in ancient Hellas with its history of time-honoured cultivation, perhaps for millennia. Here are two extremes in the agrarian structure, the self-grown city, sometimes with surrounding villages; and farms with individual plots of land and agricultural buildings as separate entities. It is clear that in the latter case the facilities for intensive cultivation are far more favourable, for instance, because transport to and from the tilled soil is minimal. Once settlement is concentrated in a town or in a village, a periphery will necessarily develop where the outlying fields are accessible only with difficulty and usually cultivated less intensively. Consequently the yield is smaller and the risk of waste during transport or owing to hostile interference is greater. Widely dispersed agrarian activity therefore constitutes a risk, and in view of the unrest that persisted between the city-states of Archaic and Classical Greece, it is understandable that fortified cities were often preferred. Examples of short raids are manifold – we shall confine ourselves to the episode mentioned by Herodotus (3.58) concerning the Samian raid against Siphnos: the Samians came ashore, ravaged the countryside districts and extorted 100 talents from the Siphnians, many of whom were prevented from retreating to their city. The situation appears to be typical for the Archaic and Classical periods. As is to be expected, Herodotus does not inform us whether Siphnians were resident outside the city. A considerable amount of material on the destruction of agriculture owing to warfare has been collected and discussed by V.D. Hanson (1983).

Whereas traditional archaeology has rarely taken an interest in peripheral dwellings, this has become the centre of interest of landscape archaeology which has come to play a prominent role during recent decades. Intensive studies of the Greek historical landscape have brought much new information to our attention, and we encounter an ever more refined methodological approach. Whereas, in earlier days, one plotted the so-called 'sites', frequently readily identified as farmsteads, caution now prevails, and references are made to greater or smaller concentrations of artefacts. Interpretations vary, and here we encounter a well-known phenomenon: the more sophisticated scientific methods become, the less inclined will the scholar be towards drawing a clear and unambiguous conclusion.[53]

53 See the general discussion by Snodgrass in Bintliff/Snodgrass 1985.

AGRICULTURAL BUILDING

It is to Osborne's credit that he aims at a comparison between various areas that have been examined by surveys.[54] Thus, on Melos there is a large number of ancient sites, but in the sixth century the number decreases rather significantly, whereas the size of the remaining sites appears to increase. The development continues – in the third century the number dwindled to half of what was found in the Classical period. On the northern part of Keos, on the other hand, there is a continuous growth of the number of sites, whereas the main city on Thasos throughout the entire Archaic and Classical periods seems to be the centre as far as the population goes; on the island there are very few scattered sites. On the mainland itself there are vast differences; the great number of villages in Attica may be viewed as the extreme point. It is probably too soon to undertake an accurate comparison before the final publications of a series of surveys have been made available. It would also be premature to attempt a more precise explanation of regional differences, but – with Osborne – other ways of using natural resources should be considered (such as the occurrence of metals, quarries and so on); likewise, the occurrence of fresh water is a decisive factor. Naturally, political conditions are also important. Thus, in Sicily, we find that a series of small Archaic sites vanish in the course of the fifth century: no doubt this has to do with the unstable conditions on the island. Examinations, mostly in the territory of Gela, show that the territory was re-populated in the second half of the fourth century when political stability had been re-established. The case of Metapontum has already been mentioned. It has been calculated that there were approximately 700 individual farms distributed over an area of about 6,500 ha. We cannot of course be sure that they were all populated at the same time, but nevertheless we have here a high degree of de-centralized habitation.[55]

Finally, local differences deserve to be pointed out with yet another example, Chios.[56] From this island there is no systematic survey, but there are sufficient data to form a picture. The main city of Chios is to be found under the present city of the same name, in the centre of a narrow plain which is very fertile and, in more recent time, intensively cultivated.[57] For this reason there is little archaeological information to be obtained, but we may be sure that this plain, Kampos, was the most important agrarian area of the island in antiquity too. Besides, there are several other major structures, Emporion being the best known and one of the oldest Greek villages to be excavated.[58] It seems to have been abandoned as early as c. 600, replaced by rather more scattered building structures. According to the excavator it is unlikely to have housed more than about 500 people in all and was, in its

54 1987, 56 ff.
55 Coarelli 1981.
56 Yalouris 1986.
57 Bouras 1984.
58 Boardman 1967.

essential features, agrarian, as may be concluded from remains of terraced agriculture on the surrounding slopes. From the second half of the fifth century there are outlying sites which are often interpreted as farms. Boardman has excavated two of them, one near Emporion where the *trapetum orbis* mentioned above was found, and one near the small village of Delphinion north of the city of Chios.[59] This place plays a part in the Peloponnesian War (Thucydides 8.38 and 40). Neither excavation provides a clear picture of the farm buildings, but they both show clear evidence of terraces carefully laid out; upon them elements of buildings were placed. Besides, on the plateau to the north-west of Kampos a number of farmsteads have been found. The plateau is at a height of 300–500 m above sea-level, and the rock rises steeply upwards to the plateau from the plain. The area is very barren and full of loose stones. By extreme exertion, in antiquity, stones were cleared away and arranged in enormous piles, one of which is 7 m in height and approximately 40 m in diameter. In this way, firm ground was reached, suitable for laying stone foundations for buildings made of sun-dried clay. In one case, a very large water tank was uncovered.[60]

Today, the area is largely desolate. In one place there are goat-folds and a hut; it is not difficult to imagine that the ancient farmer lived a life much like that of today's peasant with an emphasis on sheep and goats, perhaps all year round, as we still see it on the barren northern side of Chios. Yet it cannot be excluded that crops of various descriptions may have been cultivated.

The buildings in this inhospitable part of the island have led the editor to suggest that those buildings were erected by the survivors of a Chiotic uprising of slaves in the late fifth century. This may be a possibility, but habitation appears to have continued for several centuries, so we are not dealing with temporary settlements. However this may be, it is a striking example of the very considerable regional differences within a comparatively small area, and it goes to show that even extremely infertile districts have made survival possible. This is a case of exploitation of the extreme marginal soil which, in antiquity, is often called *eschatia*.[61]

The example of Chios shows how useful a combination between surveying and excavation may be. Landscape archaeology furnishes a series of general features in the history of larger areas; but it is only when proper soundings and excavations can be undertaken that we may arrive at a reliable chronology and an accurate definition of the purposes of population and their buildings. Therefore, it would be wrong to say that landscape archaeology has outwitted traditional archaeology – rather, it is an essential, indeed indispensable supplement to the former. Even when excavations of isolated buildings in the countryside are carried out, we cannot be sure that we obtain results which in any definite way throw light on agriculture. A

59 Boardman 1958–9 and 1956.
60 Lambrinoudakis 1986.
61 See the discussion by Lewis 1973, 210 ff.

couple of examples of buildings excavated in this manner may serve to illustrate this point of view.

The country house near the Cave of Pan at Vari[62] is situated on the slope of Hymettos, facing south. The floor in the northernmost rooms lies mostly on the top of hewn rock whereas the southern rooms rest on an artificial terrace. The house (Figure 4.1) measures 17.6 m by 13.7 m and comprises, approximately, an area of 205 m², of which no less than 117 m² is occupied by a courtyard with columns. Only the bases are preserved, and in the opinion of the excavators the columns were wooden. In front of the main entrance to the house, at its south side, there was what the excavators have termed a 'veranda'. The house was surrounded by an irregularly built wall. Only the lower parts of the walls are preserved and the walls of the house

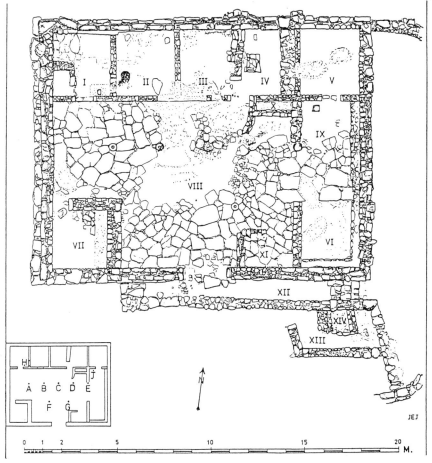

Figure 4.1 Country house near the Cave of Pan at Vari (Jones/Graham/Sackett 1973)

62 Jones/Graham/Sackett 1973.

seem to be a socle for a mud-brick superstructure. Presumably, the roof was made of tiles, but few fragments have been preserved. We do know, from other sources, that when abandoning a house you took with you woodwork like doors, sometimes window-frames, shutters and the like and sometimes even the roof tiles.[63] This material could obviously be used elsewhere. It is quite clear that a house, thus deprived of vital architectural parts, would soon fall into decay once the mud bricks were exposed to wind and weather. Findings of pottery show that the earliest material belongs to the end of the fifth century, and probably the house has been in use for a couple of generations, then abandoned, but later on re-used from time to time.

Practically all the rooms face the central courtyard. At the rear there are four rooms of approximately the same size, and to the east there are two rooms, one connected with a back-room (V). Along the eastern part of the wall and in Room IV there is a low bench. Only Room VII, at the south-west corner, is of a more solid construction, and the excavators suggest there may have been a tower here. No findings suggest that the house was in any way associated with agriculture, but in our opinion these circumstances cannot be decisive. It seems that the house was abandoned deliberately, and as much as possible was taken away, perhaps including a wooden press. If the dating to the end of the fifth century is to be maintained, we cannot expect more solid machinery for the production of wine and olive-oil, and as far as threshing-floors are concerned, only a few are preserved from the Classical period and their absence in this particular case bears no evidence. The location of the building makes it almost mandatory that the occupants must have had some connection with agriculture, and the excavators refer to nearby terraces where cultivation was performed. Besides, the typical terracotta beehives, which are known from other sites, have been uncovered, and in their report the excavators include an important chapter on ancient apiculture.

This is an exciting example of a building whose location indicates an association with agriculture, whereas the findings give no clue in that direction. So we may say that as evidence of agriculture this house stands out thanks to the absence of archaeological discoveries! This absence may be interpreted as due to the absence of agriculture, an extremely efficient evacuation of the house, or a combination of these two possibilities – that the type of agriculture connected with the building was so poor in implements, or indeed devoid of tools, that all vestiges could easily have been removed. Here we must point to the possibility that whatever may have been left behind could easily have disappeared in connection with re-use of the house in later periods.

The small country house near Vari[64] is a much smaller house lying in the open country, recently excavated and not far from the former. Here we find

---

63 Thucydides 2.14.
64 Lauter 1980, 242 ff.

a house situated above the fertile plain on the calcareous rock itself. The area of the house is 13.07 by 9.65 m, occupying in all 142.5 m² out of which the central open courtyard takes up a considerable part of the entire structure (Figure 4.2). Some fragments of pottery would seem to indicate that this

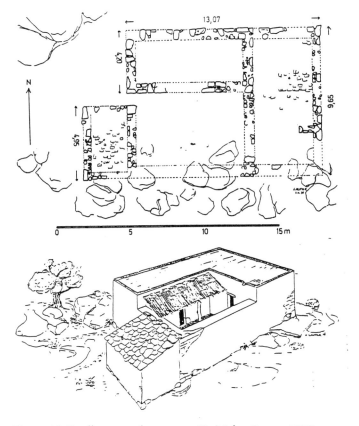

Figure 4.2 Small country house near Vari (after Lauter 1980)

building may be dated to the same period, at least approximately, but here again there are no remains to prove that we are dealing with a farm building. It is merely the location that points in that direction. In the opinion of the excavators other, but even simpler, parts of the building may have been placed on a flat plateau immediately to the west of the house, but no traces of any such remain. It is, of course, possible that the plateau may have served as an extension of the courtyard outside the living quarters of the house itself, but at the same time we cannot be sure that the house was designed for residence all the year round. So far, this is the smallest of known country houses situated in the open countryside.

We could site further examples of country houses from the Classical period where no trace of agriculture was found, but it would not change the situation and would in fact tell us little about country life.[65] From the chora of Metapontum several farmsteads have been excavated, but as yet they have not been published in a way that would bring us further in our present discussion.[66] It is not until we arrive at the Hellenistic period proper that complex buildings constitute clear evidence of their functions. The greater part of the material hails from the excavations, previously mentioned, undertaken at the Crimean Chersonesos. The excavations have been undertaken over several stages, but we shall select a well-preserved country house from the second century, namely Strzheletskii No. 26, most fully described in the article by Maria Dufková and Jan Pečírka.[67] The excavator who has lent his name to the numbering appears to have published his findings with a thorough commentary, also with reference to Roman agronomists. We have already warned against the danger of using Italian sources, but we admit that we are unable to offer a qualified criticism or evaluation as long as the discussion is available only in a very restricted version for the benefit of scholars in western countries.

As for Farm No. 26 (Figure 4.3) it was rebuilt after a fire, probably in the second century. Its predecessor was much smaller, as far as it may be reconstructed from the strata below the present one, and it is scarcely worth our while to discuss the first building. In part, it is to the south of the present building. Rooms 1–5 are essentially designed for agricultural purposes, and the tower appears to have served as a habitation for the vilicus (*sic*!). It is particularly Room 3 which is clearly identifiable, containing two stone press-beds and a container lowered into the floor. Here the grapes were pressed and the wine perhaps stored in Room 2 where many remains of *pithoi* and *amphorai* were uncovered. Rooms 6–8 are for occupation, Room 9 perhaps a kitchen for the slaves.

> Since a farmhouse must have cowsheds, Strzheletskii sees them in the rooms 10 and 11 by the wide gates leading to this part of the inner yard from the outside. Room 12 with a floor above it may have been a summer shelter for the stock.[68]

It seems as if the authors are not entirely convinced by the interpretation, and there is no information to show whether there are specific conditions to indicate the permanent presence of cattle; but as we have stated, at this point we cannot exercise any constructive criticism. The building measures 26 by 22 m, and the thickness of the walls would indicate that large parts have had an upper storey. This was an impressive building with rooms centred round

---

65 The Athenian country houses are easily found in Osborne 1985, Appendix A. A recently excavated building from Messenia is published by Kaltsas 1985; see further Lohmann in Wells, ed., 1992.
66 Plans of a few country houses are in Adamesteanu 1974, 83–4.
67 1970, 167 ff. Skydsgaard tried in vain to visit the area in 1990.
68 Ibid. 171.

AGRICULTURAL BUILDING

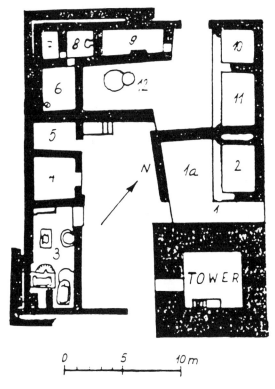

Figure 4.3 Country house no. 26 at the Crimean Chersonesos (Dufková/Pečírka 1970)

a courtyard; from this, brick stones in the north-west corner seem to have continued on to a wooden staircase.

This entire building leaves you with an impression of an extremely effective and technical construction, something unknown from the outlying farmhouses of the Archaic and Classical periods. This may be due to coincidences, but the farmstead is typical for Chersonesos. In the pre-Hellenistic period much simpler implements were in use, and if we can rely on vase paintings, the pressing of grapes may often have taken place in the vineyard, therefore leaving no traces in the farmyard. If the two older buildings near Vari were in fact country farms, they testify to a type of agriculture infinitely less sophisticated than that which we find in the Hellenistic and Roman periods.

Thanks to the specific conditions associated with the planning at Chersonesos, it is possible to determine quite accurately the acreage and crops of the farm. Like the other farms, No. 26 is located on a piece of land surrounded by a wall, altogether an area comprising 29 ha. The area is terraced and divided into smaller fields. Thorough palaeobotanical investigation makes it possible to conclude that grain was grown over an area of approximately 14 acres, vine about 53 acres, and fruit-trees about 2.5 acres. The farm building itself was limited to no more than just over 0.22 ha. So we are faced with a highly specialized type of agriculture with its emphasis on the production of wine, no doubt meant for marketing. This is typical for the area. Excavations have left us with a series of unique niceties of cultivation, such as plant pits hewn into the rock for the roots of the trees, extensive hewn trenches filled with earth and sometimes with drains consisting of pebbles at the bottom and walls built to support the soil so that the trees would stand in raised beds. It is supposed, furthermore, that a small area can be identified with a nursery designed to develop a mass production of cuttings, in other words a highly advanced production potential which we have not as yet encountered on native Greek soil. This must have to do with very specific marketing structures, but it is not our task to discuss this matter here. The contrast with agrarian conditions in the mother-country is, however, striking.

It is characteristic that the compact piece of land belonging to the individual farms at Chersonesos is of considerable size. Many estates have an adjoining piece of land measuring up to 30 ha., some slightly more, some slightly less. If we compare the chora of Metapontum, the number of farms indicates that there we find farms of a much smaller scale, perhaps approximately 9 ha. on an average, but as the sites are unevenly distributed in the area, the individual farms were probably of very varying acreage, and we may assume that changes in property rights occurred during the period when the area was so closely populated.

If we look at the literary tradition, which – as is to be expected – provides us with examples primarily from Attica, it is noteworthy that the largest areas specifically mentioned are 300 plethra, that is approximately 27 ha. This is the ancestral estate of Alcibiades in Erchia (Plato, *Alc.* 1.123 c) and a similar estate purchased by a thrifty person (Lysias 19.29). In both cases we are given round figures designed to illustrate what may be regarded as sizeable farms. Furthermore, as an example of the moderation of earlier generations, Demosthenes tells us that Lysimachos, son of Aristeides, was presented with 100 plethra of farmland and 100 plethra of planted land on Euboea – all in all about 18 ha. There is not much to indicate that Lysimachos would have moved to Euboea for this reason. Finally, Isaeus mentions a couple of small houses and an area of 60 plethra 'on the plain', about 5.5 ha. In the context this is looked upon as an insignificant piece of property, but this, of course, must be viewed in connection with what the opposing party has derived from the estate in question.[69] And there is the 'farm of Phainippos' (pseudo-Demosthenes 42), frequently mentioned; by

---

69 The passages are often discussed, see Finley 1973c, 56 ff., Davies 1981, 52 ff.

means of de Ste Croix's analysis (1966) it has dwindled considerably. It should be noted that this farm is defined as *eschatia*, that is to say, it was located away from the territory normally cultivated, and no one can tell how large it in fact was; and it would appear that the speaker has no wish that the judges should know. They are merely to form an impression that it is huge. When, in legal documents, the location of an estate is to be established, the neighbours are often mentioned by name, and cases of controversies between neighbours are well known from the speeches. It shows how close the farms were to one another in what was essentially the agricultural territory.

The indications of acreages as given above are by no means official figures. If, then, we turn to a considerable number of evaluations and sales, we are on safer ground, but here the difficulty is in converting amounts to area. Pritchett has collected a large amount of material in his commentary to the Attic stelai, but wisely refrains from making deductions from amount to acreage.[70] In that connection the quality and location as well as the state of cultivation at any given time will play their part, and the sources are silent on these matters. Likewise, we must disregard the leasing contracts which are preserved, because although they do mention rent, the size of the area leased is not indicated. It is, however, generally agreed that most Attic farms had only a modest acreage, if only for the very reason that so many citizens did in fact possess land. Dionysius of Halicarnassus mentions a bill proposed in the year 403 by a certain Phormisios; he suggested that the number of the citizens be limited to those who owned land, which would entail the loss of citizenship for 5,000 persons (*Argumentum ad Lysiam* 34). So, the remainder of the citizens were landowners, but how large this remainder was is a matter of considerable controversy. Peter Garnsey (1985) and others tend to diminish the number of citizens, at the same time increasing the cultivated area, whereas, based on other points of view, M.H. Hansen, for example, contests these calculations (1988). However interesting this discussion may be, it should not veil the fact that our information with regard to the size of the individual farms in Attica, as well as in the rest of Greece, is extremely limited.[71] Attempts at converting the Solonian *tele* into areas yield results so different that caution must be recommended.[72] We shall prefer to follow Alison Burford Cooper (1977-8) who estimates that the *zeugitai*, who were admitted to serve in the army as hoplites, are not likely to have owned land less than about 4-5 ha. The fact that the size of individual farms was modest need not necessarily tell us much about the actual rights of ownership; the same person may well have owned several farms and thereby accumulated wealth in terms of land.[73] This is of interest in so far as the economy is

70 Pritchett 1956.
71 As to the Helots of Sparta we do not know anything concerning their plots; see Cartledge 1979, 165 ff., and the reflections by Jameson in Wells, ed., 1992.
72 Skydsgaard 1988a.
73 Davies 1981, Osborne 1985.

concerned, but scarcely with regard to agricultural technology. Amalgamation of smaller farms into larger entities does not involve great advantages in a sub-technological society. This is generally valid and has also been made applicable to the interpretation of Roman agriculture where the latifundia, to an increasing degree in research, have been replaced by assemblages of *villae rusticae* under the same owner.[74] It is, therefore, a matter of wonder that Lauter (1980) speaks of 'latifundienartige Güter' when referring to 'das Marmorgut bei Porto Lombardo'. Generally speaking, it is dangerous to draw conclusions from the size of a farmstead and the quality of the building to the area under cultivation. For this, there are far too many elements of uncertainty.

Finally, it should be emphasized that the numerous plots of land mentioned in inscriptions are not necessarily independent farms. Quite often, a plot of land may have been cultivated from a distance, and Greek terminology does not show whether independent farms are involved. The mere fact that, as mentioned above, presses are often found in urban or built-up areas is sufficient to show that part of the agrarian population were indeed residents of some town or city – people who, like Euphiletos in the first speech of Lysias, went off to their own piece of land in order to cultivate it. Pritchett has submitted a thorough analysis of the relevant Greek terminology, and we shall reproduce the categories proposed by him in the Attic stelai:[75]

1	*agros*	field for cultivation in the country
2	*ge psile*	land cultivated for cereals, vines and the like
3	*gepedon*	plot of ground (exact meaning uncertain)
4	*dryinon*	oak grove
5	*kepos*	garden (non-specialized)
6	*oikia*	private residence
7	*oikopedon*	house-site
8	*orgas*	woody mountainous tract
9	*pityinon*	pine grove
10	*synoikia*	tenement-house
11	*chorion*	land, landed property

We cannot make a very general statement about the degree of parcelling-out of the land of the individual estates. We have already mentioned the high degree of variation in the patterns of habitation – with a structure like that of Thasos, with one dominant urban population, the possibility that the land has been parcelled out in small plots is much greater than in the case of northern Keos, where the habitations seem to have been more widely spread. Later (p. 127) we shall see how, by way of legislation, attempts appear

---

74 White 1967a, Skydsgaard 1969.
75 1956, 269.

to have been made in order to counteract the nearly complete disruption into small lots that is characteristic of many modern Greek agricultural communities. It is, nevertheless, worth stressing that there are certain advantages connected with such scattered lots in a sub-technological system of agriculture because it implies a 'division of risks'. In his classical analysis of a village on the peninsula of Methana, Forbes (1976a) has shown how the widely spread sites ensure the owner a reasonable and stable average crop in an area described as being extremely varied indeed. To this we may add that a forest fire like the one that laid waste large parts of the plantations round the village a few years ago will not ruin a minority, which loses everything, but affect many among whom everyone loses something. It is only with the industrialization of agriculture in the most recent past that the considerable parcelling-out of arable land has turned out to be a definite impediment for further development.

Finally, we shall have to mention a problem which is of importance for our evaluation of the Greek agricultural potential in antiquity, namely, the use of terraced cultivation. Anyone taking a walk in the Greek countryside will be struck by the astonishing ability the local population shows in expanding the arable land by means of terraces for cultivation; sometimes they characterize an entire landscape. Often small terraces can be found far off the beaten track and away from built-up areas, testifying to the fact that here, too, is a niche that has been utilized. On several occasions we have mentioned terraces in connection with excavated country houses, in Attica, on Chios and elsewhere. It is one of John Bradford's merits to have shown the existence of abandoned terraces in Attica and on Rhodes by means of aerial photography, subsequently – aided by surveys of the site – rendering it likely that they hail from ancient times.[76] Unfortunately there is no unambiguous ancient Greek term for installations of this kind. Jameson suggests that the word *haimasia*, *Od.* 18.357, should be interpreted as a terrace.[77] Eurymachos challenges Odysseus, who is dressed like a beggar, by saying, 'Stranger, wilt thou serve like those who in a field of an *eschatia* gather stones (*haimasias legon*) and cultivate the large trees?' *Haimasias*, in the lexicon, is translated as 'wall of dry stones', although we cannot say whether they were put to use as supporting walls for the soil on slanting ground, or as markings round a country estate. The context of the *eschatia* indicates that we are not within the territory which is customarily cultivated. Nor does the parallel passage mentioned permit a safe interpretation (see Menander, *Dyskolos*, 377), whereas the usage in *Od.* 24.224, seems to indicate a simple wall round the garden of Laertes. Once again we have to admit that the Greek vocabulary is not very specific and does not lend itself to any indisputable interpretation. Nevertheless, the fact that terraces were placed,

76 Bradford 1956, 1957.
77 Jameson 1977/8, 128, note 32. A preliminary study on terracing will be published by Rackham and Moody in Wells, ed., 1992.

with a perfect *al secco* technique, to support the country houses, and the presence of unquestionable traces of cultivated terraces which do not seem to have been in use since antiquity, compel us to assume that a sophisticated and refined technique of terracing with a view to extending the arable area was brought to bear.

Summing up, we may conclude that the various written and archaeological sources concerning agricultural settlement leave us with an impression of enormous differences and an untold number of possibilities in the way of interpretation, none of which makes it feasible to point to any one feature as being the norm. The peasants endeavoured to adjust to conditions offered by nature as well as they could, and evidently understood to take advantage of the available niches. By itself it is a step forward to be able to ascertain all this, although it is scarcely sufficient for someone who wishes to write economic history.

5

ANIMAL HUSBANDRY

Our knowledge of Greek animal husbandry is somewhat limited. The basic literature on this topic is still O. Keller, *Die antike Tierwelt* (1909) and a series of often admirable articles on individual domesticated animals in *Pauly–Wissowa's Realencyclopaedie*.[78] Furthermore, there are German dissertations from the University of Giessen which deal with the most important domesticated animals: thus A. Hörnshemeyer, *Die Pferdezucht im klassischen Altertum* (1929), K. Winkelstern, *Die Schweinezucht im klassischen Altertum* (1933), Otto Brendel, *Die Schafzucht im alten Griechenland* (1934), and K. Zeissig, *Die Rinderzucht im alten Griechenland* (1934), all written by authors with practical experience in agriculture. Today they may appear slightly outdated. Naturally, we find sections in works of a more general nature which deal with the subject (quoted on p. 19), and in particular we should mention W. Richter's treatment in *Die Landwirtschaft im homerischen Zeitalter* (1968), which contains very useful notes and references. If, on the other hand, the reader wants to obtain a general view of cattle-breeding and its relation to agriculture, it is not so easy. The topic was discussed at the Ninth International Congress for Economic History in Bern (1986); the contributions were published in 1988 under the title *Pastoral Economies in Classical Antiquity*.[79] On this occasion it became clear that there were essential disagreements among the participants, not least within the Greek zone.

It is not difficult to find the reason for varying interpretations. Whereas, in different ways, we can form an opinion of domestic animals and the species to which they belong, as well as of their appearance, particularly thanks to numerous representations in art,[80] it is much more difficult to arrive at a proper understanding of the role animal husbandry played in Greek agriculture. We have already observed that the few farms excavated

---

78 Steier 1938, Olck 1907, Kraemer 1940, Orth 1921a, Orth 1921b, Richter 1972, Orth 1910, Orth 1913.
79 Whittaker 1988.
80 Richter 1930; Kozloff 1981; Kozloff/Mitten/Sguaitamatti 1986; Bevan 1986.

furnish no information regarding stables, and so on, and our main sources on agricultural matters, Hesiod and Xenophon, hardly ever mention livestock. Hesiod confines himself to mentioning the draught-oxen (*Works and Days*, ll. 405, 436 and 606) where winter fodder is collected for them and for the mules. Sheep, goats and oxen freezing in winter are mentioned; sheep, however, are protected against the cold by their coat (ll. 515 ff.). We are also told the time when sheep, boar and bull should be castrated (l. 786), and when a heifer should be sacrificed (l. 590). It seems evident that especially the draught-animals are of interest, and it may be noted that you shall purchase one ox (l. 405); elsewhere Hesiod mentions the plough team in the dual (l. 437) with the epithet '9-year old'.

Xenophon is even more silent. Ischomachos buys his horse, and instructions are given concerning the purchase of a riding horse in *De Re Equestri*. Apart from that, the draught-animals appear especially in connection with threshing; these animals (*hypozygia*) are oxen, mules and horses (*Oeconomicus* 18.3). It may be remembered that burning the stubble after harvest is recommended so this does not serve for grazing during the fallow period.

Deducing from this information that there were no domestic animals would be erroneous, and a closer perusal of both authors will show that animals do appear in other contexts. Hesiod is herding sheep when the Muses approach him (*Theogony*, l. 22); and through work you become rich, the expressions used being *polymelos* and *aphneios*, i.e., possessing many sheep and being rich (l. 308). In the same way, the sheep recur in the somewhat more philosophical discourse in *Oeconomicus* 1.9: only if you are successful in sheep-breeding, the animals constitute wealth; likewise, cattle-breeding or rather sheep-breeding (*probabeutike techne*) is directly associated with agriculture (*georgia*), the former giving to man material to be sacrificed to the gods in order to please them, as well as something which benefits man himself. Here, we may think of the sheep's wool and milk, etc., or in a wider sense the entire stock-breeding, although the expression is vaguely naïve. In the philosophical writings too, Xenophon mentions sheep as a natural element, but usually in contexts so general that we cannot determine what the conditions of the livestock were and the relation to farming.

The main source for Greek cattle-breeding is the *Historia Animalium* by Aristotle. Thus, we find ourselves in the same situation as that which applies to agriculture; the main purpose of the source is not an attempt to describe cattle-breeding as such, but rather to present a classification and a description. But here and there, scattered throughout the books, there are hints which reflect how the author observed the life of domesticated animals. The modern reader cannot but admire the enormous amount of empirical material collected in this work, but at the same time it must be emphasized that Aristotle does not take a particular interest in these animals. One might have thought that the task of describing them was too easy for him. In fact, we often know more about anatomical details than we do about everyday happenings.

The sixth book is the most profitable part (571 ff.). This contains an account of the rut and heat, mating, the food and the rearing of the young ones. From this it appears that Aristotle takes it for granted that most domestic animals live in herds. In the period of heat, one stallion is calculated to cover some 30 mares, and the stallions will compete among themselves; likewise, the bulls will often graze by themselves until the period of mating when they will go in search for the cows. As a matter of curiosity it is mentioned that Epirote bulls may be completely out of sight for three months of the year, a circumstance which presupposes the existence of very large areas for the use of the cattle there. Also rams and billy-goats are aggressive during their period of heat, whereas dogs and pigs will mate any time of the year. These two species are known as *synanthropeuomena* which should probably not be taken to indicate domesticated animals in general – here the adjective *hemeros* is normally used – but as a designation for animals which live together with man. Dog and pig are listed in the same category again (542a) where domestic birds with several broods are added; the same terminology is also used about insects which are able to winter in human dwellings (599a). It seems as if Aristotle makes a distinction between animals living in herds and animals living in the close vicinity of people. We stress these statements because, in the context, they express a simple and evident observation and not an attempt at systematizing.

Domestic animals were originally wild species tamed by man. Aristotle is fully aware of this (488a). The process seems to be as old as the domestication of cultured plants, and we shall not here pursue this process in detail. Like cultivated plants, domesticated animals have developed characteristic species by means of more or less deliberate breeding. Aristotle is conscious of this too, and we find many suggestions as to how breeding animals should be selected. We cannot follow the development of the domesticated animals in detail, nor is this necessary for our purpose. Several of the animals in question probably never existed in their wild form in Greece but were originally imported – the last of them perhaps the horse, which arrived in the Bronze Age – but all the animals are mentioned in the Mycenaean tablets, which testify to a large animal husbandry. However, we shall not pursue this matter here, but review, however briefly, the most important domestic animals with an emphasis on the description given by Aristotle.

HORSES, DONKEYS AND MULES

The horse was definitely a luxury and a status symbol. Several designations of the upper classes in the Greek city-states point in this direction – *hippobotai* on Euboea and *hippeis* in Attica. Lefebvre des Noëttes (1931) has shown clearly that ancient harness made the horse unsuitable for dragging heavy loads. One may wonder at this, but the explanation is probably quite simple. Horses were not required for heavy work, which was carried out by oxen, donkeys and mules – animals that require much less fodder. So the Greek horses were first and foremost mounts and race-horses. We find them richly represented in Greek vase painting from the time of the geometric style, especially in connection with funerary ceremonies. The Homeric battle-chariot which conveys the warrior to the battlefield, whereupon he fights on foot, seems to be a specifically epic and literary phenomenon. It is probably under the influence of these patterns that we find, in later vase paintings, quite frequently, heavily armed men with a chariot drawn by horses as a variant of the theme which has been called 'Kriegers Ausfahrt'. In real life the part played by the horse is much more modest, and it is doubtful whether it found much use in agriculture. On the other hand, a number of states do have a light cavalry[81] (Plate 5.1).

Plate 5.1 Athenian horsemen, Elgin Marbles, British Museum

In his description of the horse, Aristotle (575b) allots a period of 18 to 20 years as its normal lifetime. Horses are sexually mature at the age of 2, but older breeders are preferable, and it is only after the shedding of the last teeth, when the animal has reached the age of about 4½ years, that it is fully developed. The mare foals normally after a pregnancy of just over 11

81 Bugh 1988.

months, and Aristotle stresses the fact that the mare should foal only every other year 'so that she may as it were lie fallow'. It is better that she should be made in foal only every fourth or fifth year. Therefore, this is a case of a slow and quite costly reproduction, but then, the horse is procreative throughout its life.

While the horse demands a considerable amount of fodder, the donkey is much more easily satisfied (Plates 5.2, 5.3). Both are included among the graminivorous species (*karpophagoi kai poephagoi*, 595b), but the donkey has a longer span of life, often more than 30 years. Like the horse, the donkey's pregnancy is 11 months, and it gives birth in the twelfth. Unlike the horse it seems that the she-ass was mated in the first heat immediately after (577a). Twin-birth is rare. Not only then is maintenance of the donkey cheaper, but reproduction takes place much more frequently. The mule is the sterile crossbreed between the male donkey and the horse mare. This animal combines the power of its mother with the temperance of its father and is, therefore, a valued domestic animal. Conversely, the hinny, the crossbreed between the she-ass and the stallion, in our time is looked upon as far inferior, but Aristotle refrains from discussing this matter. Although he knows both forms of crossbreeds, all he says is that the offspring will take after the mother. The pregnancy of the donkey is said to be the same as that of the horse, and it is emphasized that the same mare should not be continuously used for breeding of mules as she will, in that case, become sterile (577b, ff.). The creation of crossbreeds is not quite without problems because the male donkey will not spontaneously serve the mare. The male donkey must have suckled a horse mare, and such male donkeys are called *hippothelai*. They serve the horse mare with as much eagerness as does the horse stallion. The mating, likewise, takes place on free land (*en te nome*). In other words, the breeding of mules is handled under extensive forms provided you are in possession of the proper breeders, and you would scarcely maintain a *hippotheles* without having access to several mares, particularly since the same mare should not constantly be used for crossbreeding. The remark that a donkey mating a pregnant horse mare leads to an abortion would seem to indicate that in stud farms mules could also be bred. It seems reasonable to assume that the horse mare should also 'lie fallow' after foaling with a mule, and this means that the reproduction of mules is slow and therefore costly. At the same time we are reminded that the mule has a long span of life. Aristotle does not indicate any normal duration of life, but mentions a single case where a mule lived to an age of more than 80 years. This was looked upon as something quite out of the ordinary.

Although Aristotle is well aware of the phenomenon of castration (631b), he does not specifically mention castration of the equine domestic animals. Therefore we cannot be sure whether breeding takes place also in connection with very limited husbandry. Aristotle mentions that horses reared privately live longer than those from large establishments but he does not say what is

Plate 5.2 Mating donkeys, Attic red-figure oinochoe, Munich

Plate 5.3 Donkey with packsaddle, cameo, Thorvaldsen Museum, Copenhagen

normal practice. We should remember that Xenophon has no intention of buying castrated stallions for riding, but, as we have mentioned, he does purchase the mount. This is not a case of home breeding. Presumably, we have to assume that normally the farmer would have to buy at least the mule and perhaps also some of his donkeys.

OXEN

Cows have a life span of about 15 years, the bullock, the castrated bull, about the same, whereas bulls can live somewhat longer (575a). If a bullock has been trained to lead a herd, it will live longer than the 15 years owing to a greater quantity of fodder and because it is not a beast of burden. Sexual maturity occurs during the second year of the animal's life, mostly after the twentieth month or at the age of 2. The pregnancy is 9 months, and the cow may be served shortly after calving. Apparently there are no limitations with regard to the frequency of pregnancies, and the animals are procreative throughout their lives. Normally one calf is born, rarely two. This means that one cow is able to produce more than ten calves – in other words, a reproduction much larger than that of the horse – but not much larger than that of the donkey. The ox is fully mature at the age of 5. Here, Aristotle refers to the Homeric usage *pentaeteros* and *enneoros*, words which have the same meaning. In that case, the latter must indicate an age of nine half-years. Others feel that the juxtaposition of the two words, as Aristotle has them, would indicate that the ox is unimpaired from the fifth to the ninth year. We have already seen that Hesiod calls the plough-oxen '9-year olds', but West is undoubtedly right in his interpretation when he calls it 'a formulaic age'.[82] Castration of the bulls takes place when they reach the age of one year; otherwise they would not grow on satisfactorily. The operation is described in detail (632a), and we must assume that it is well known and commonly practised.

The function of the ox is primarily its performance as a draught-animal, designed for strenuous labour such as ploughing and heavy transport (see Plate 3.3). We have seen representations of mules as a plough-team, but the ox is shown more often. Although the bullock is the stronger, the cow is also used as a draught-animal in many civilizations; from the references in our texts we cannot determine the sex of the animals. Hard labour, however, diminishes the fertility of cows considerably, and it is probably safe to assume that most draught-animals were bullocks. The observation concerning the longer life-span of the leading bullock on the pasture, together with other indications adduced above, would suggest that breeding normally is extensive.

Apart from being used for labour, the ox delivers milk, but the Greeks

---

82 West 1978, 269. The two different interpretations are to be found in the Loeb edition and in the Budé edition.

Plate 5.4 Herakles leading a bull, Attic red-figure amphora, Boston (after Pfuhl)

Plate 5.5 Cows in sacrificial procession, Elgin Marbles, British Museum

were not milk-drinkers; besides, they preferred milk from sheep and goats for making their cheese. Finally, the ox is an important sacrificial animal, its meat a favourite dish and its hide was used for a multitude of purposes (Plates 5.4, 5.5). Naturally, these aspects are of no interest to Aristotle, the zoologist, and consequently they are not mentioned.

SHEEP AND GOATS

According to Aristotle, the sheep's lifespan is 10 years and the goat's 8, but few are allowed to live as long as that. The castrated leading ram may live to be 15 years (573b). The ewe's pregnancy lasts 5 months, and mostly she lambs twins or more. Normally they lamb only once a year, but under favourable conditions they may lamb twice. Sheep and goats are sexually mature when they are a year old, which of course yields a fairly rapid reproduction when compared with equines and cattle. The sheep are looked upon as the most stupid of animals, and shepherds often have to bring them in for the winter as otherwise they would freeze to death in the snow (610b). They willingly follow the castrated leading ram, whereas, during grazing, the goats soon spread. Both animals are regarded as graminivorous, the sheep grazing to the naked soil, the goats mostly nibbling the fresh shoots (596a). It seems clear that Aristotle thinks of them as animals living in herds, and the shepherds are mentioned frequently, as are the dogs to which a separate section is devoted (Plates 5.6, 5.7). Sheep and goat are milked. Naturally this is beyond Aristotle's sphere of interest, but he does have an excellent section on milk (521b, ff.) indicating how cheese is made by adding fig-juice or rennet found in the stomachs of suckling animals. We shall not enter into a detailed discussion of cheese-making, but merely observe that cheese was part of the normal diet as a natural supplement. Cheese was made near the place where milking was done so that transport was avoided, milk being perishable, especially in a hot climate. Cheese-making requires few tools: facilities for warming the milk, rennet and strainers so that the whey may drip from the curds which may then be pressed into moulds and cured. The entire procedure is very vividly described in the *Odyssey* (9.246 ff.). Furthermore, of course, sheep's wool was utilized. Ram and wether carry considerably more wool than the ewe. From bones recovered we are able to determine whether a stock of sheep was kept in order to produce milk, or whether wool was the main product. In the former case, bones from male animals will mostly stem from very young and not yet fully matured specimens because those are slaughtered at an early time, as they are not needed for breeding. Bones from female animals, on the other hand, will mostly be from fully grown and older specimens. If, on the other hand, you invest in the production of wool, then the bones of male animals will also stem from fully grown older animals. Unfortunately, investigation of bones from the Archaic and Classical periods has not been undertaken very

Plate 5.6 Goats and herdsman, Attic black-figure kyathos, Louvre (after Pfuhl)

Plate 5.7 Odysseus escaping under the ram, Attic black-figure lekythos, National Museum, Copenhagen

thoroughly,[83] so that as yet we cannot with any certainty determine which type of production was particularly favoured; needless to say, this may have been subject to considerable variation, and with a relatively quick reproduction rate it must have been possible to re-adjust production within a brief span of years.

We should point out that sheep and goats were sacrificial animals, and their meat was a much-coveted item of food. In view of the rate of reproduction, it was possible to set aside a fairly large number of lambs and kids, produced annually, for slaughter and offering without interfering with the size of the stock.

PIGS

The domestic pig is not looked upon as being either particularly frugivorous or graminivorous, but as a root-eating creature, *rizophagos* (595a), and by nature it is well suited to grubbing the soil. At the same time it is an animal that can be nourished by the most diverse kinds of food – in other words what we should call omnivorous. A sow may live to the age of 15 years, some even more; the pregnancy is 4 months, after which it will give birth to as many as 20 piglets. Modern experience indicates that these are exceptional cases. Aristotle does mention that the sow cannot rear very large litters of young (573a). The pig reaches sexual maturity at the age of 8 months so that the sow will normally farrow at the age of 1 year; she can then continue breeding for the rest of her life, whereas the boar breeds most favourably between his first and third year (545a). It is recommended that the boar should be fed on barley when breeding, whereas the sow should be fed on boiled barley when farrowing.

Pigs are kept almost exclusively for the sake of their meat (Plate 5.8). Therefore Aristotle has specific recipes for its fattening (595). The same applies as to other animals, you begin by starving the pigs for some days; thereupon they are given as much as they can eat. They are fattened during 60 days, but unfortunately the zoologist does not tell us at which age the fattening begins; so this bit of information is not of much use when trying to evaluate the breeding of pigs. When the root-eating habits of the animal are emphasized, the reason is probably that pigs living in the open are referred to, but as we have mentioned above, the pig is called *synanthropeuomenos*, that is, living together with man, like the dog. This does not indicate very extensive breeding.

The pig appears as a sacrificial animal in a number of cults, often not as a fully grown animal (see p. 177). It is evident that with their considerable ability in procreation, the greater part of the breeding of pigs must have been directed towards sacrificing and/or slaughtering.

---

83 Greenfield 1988, see also Clutton-Brock/Grigson 1984.

THE ART OF AGRICULTURE

Plate 5.8 Going to market with pigs, Attic red-figure pelike, Fitzwilliam Museum, Cambridge

POULTRY

The interest which Aristotle devotes to poultry is rather closely connected with egg-laying and procreation, described in the beginning of his sixth book. Among domestic birds he mentions, in particular, barnyard fowls and pigeons; geese to a lesser degree. Whereas the pigeons lay only two eggs at a time, and hatch them, the domestic hen stands out in its ability to lay eggs practically all year round. Aristotle has undertaken – or he has arranged for someone to undertake – exact investigations of the development of the embryo in the egg, and describes it in detail.[84] The fact that the greater part of the eggs are not used for hatching, but for food, is not of the same interest, and there are no precise descriptions of poultry keeping. However, there is no reason to believe that it reached the stage it did in the Hellenistic and Roman periods; the third book of Varro's *De Re Rustica* is devoted to the so-called *villaticae pastiones*. It comprises the breeding of poultry, the common dormouse, hares and rabbits, and so on, which may be kept with advantage in a limited area. Varro's terminology is frequently Greek

84 For earlier studies in the embryo see the Hippocratic *De Natura Puerorum*, 29.

(*ornithon*, *ornithoboskeion*, *chenoboskeion* and so on), but no sources allow us to decide the earlier history of this terminology. There is nothing to indicate the existence of larger and more highly developed poultry keeping in the Archaic and Classical periods. It may also be noted that the domestic hen seems to have been introduced into Greece at a fairly late date, probably from Persia. The first specific mention of the cock is found in the poems of Theognis (ll. 863 ff.), but it would be wrong to deduce, from this, that the bird was not known at an earlier date. From the Bronze Age there is only one case where bones of hens have been found, but from the time of the geometric style we have a couple of terracottas representing a cock, found in a child's grave in Attica,[85] and in later Greek art from Archaic black-figure vase paintings hens are well known, and cockfighting seems to have been a favourite sport (Plate 5.9). But it would be unwise to draw any conclusions about poultry farming on the basis of this evidence.

Plate 5.9 Cocks, Attic Tyrrhenian amphora, National Museum, Copenhagen

The goose is mentioned in the *Odyssey*; the best-known example is Penelope's flock of twenty geese which, in a dream, are killed by an eagle. They live on the farmyard and are fed by wheat thrown into water (*Od.* 19.536 ff.). Richter (1968) regards this flock as a luxury and argues that,

---

85 Coldstream 1977, 313. For birds in geometric miniature bronzes see Johansen 1982.

properly speaking, the goose cannot be considered a domestic bird in the *Odyssey*.

BEES

Honey is the most important sweetening agent in antiquity, and the production of honey played an important role. Although Aristotle displays great interest in the social structure of the bees, he does not succeed in disclosing its proper context with the one and only procreative queen bee, the workers and the drones whose only function is procreation. As we know, bees pair aloft, often at a great height, and it seems that they have not been observed and described. Therefore, the sequence on the social structure of the bees (553a, ff.) does not convey much information about apiculture, but it does reveal great interest in the subject. However, archaeological discoveries of terracotta beehives have added to our knowledge.[86] It would take us too far to engage in a more detailed discussion of this matter in our context.

In conclusion, we may say that Aristotle reveals an intimate knowledge of many aspects of animal husbandry. He does, of course, convey a number of observations that we, today, can denounce as being faulty (for instance the assertion that when sheep mate when the wind blows from the north, the offspring will be of the male sex, whereas the offspring will be female when the south wind blows (574a)). No doubt this builds on information gleaned from superstitious shepherds. He does, however, have an open eye for many details, and he mentions different races of the different animals. In our opinion, it would serve no purpose to pursue this subject further. It would require the inclusion of later sources, first and foremost the Roman agronomists who have left us a legacy of carefully worded descriptions of the individual domestic animals. These may stem from Mago of Carthage.[87] Aristotle is conscious of the importance associated with the selection of breeding animals, and we must assume that a very deliberate effort with regard to breeding has constantly taken place, including the purchase and transport of breeders aiming at an improvement of the stock. This procedure seems to have been intensified in Hellenistic and Roman times, if we rely on the exorbitant prices paid for breeding material. For this reason we shall refrain from discussing this aspect, also because anyone interested can easily find it dealt with in various other studies.

If we turn to the Homeric epics, there is an abundance of cattle. They may be cattle of the sun-god, 7 herds (*agelai*) of cows and 7 herds (*poea*) of sheep, each comprising 50 animals (*Od.* 12.127 ff.), or the cattle of the Eleans which

---

86 Jones/Graham/Sackett 1973.
87 Columella 6.1.3.

in his youth Nestor captured (*Il.* 11.678 ff.): 50 herds of cows and as many herds of sheep, pigs and goats, together with 150 mares, most of them with foal. The number of cattle which Odysseus owned on the mainland is of the same heroic dimensions, in the proud enumeration of Eumaios (*Od.* 14.100 ff.): 12 herds of cows and as many sheep, pigs and goats. The choice of words in the two passages is strikingly similar, and one wonders whether it is a coincidence that in both cases we meet persons who openly brag about achievements of their youth or about the property of their master. Besides, on Ithaka itself, Odysseus owns 11 herds of goats and all the pigs administered by Eumaios.

We also find cattle used as a unit in reckoning. It had cost 100 oxen in ransom to free Lykaon (*Il.* 21.79); Laertes bought Eurykleia for 20 oxen (*Od.* 1.431); and Glaukos exchanges his arms, worth 100 oxen, against those of Diomedes which are valued at 9 oxen (*Il.* 6.236). In these cases adjectives of the types *nekatomboios*, *enneaboios*, etc., are used, which would indicate that we are dealing with units quite commonly applied which could be used not only in the epic genre.

Finally, cattle appear in the numerous similes just as we have seen agriculture used as a point of comparison. The army of the Achaeans is likened to flies swarming round the shepherds' stable when milking takes place in spring, and their commanding officers separate their men as easily as goatherds separate their herds when they have been grazing with others in the pasture (*en to nomo*); Agamemnon towers among the people as the bull amongst the cows (*Il.* 2.469 ff.). The physician arrests the bleeding just as the rennet curdles the milk (*Il.* 5.902 f.), and Hektor lifts a stone with the same ease as a shepherd carries the wool from a male lamb (*Il.* 12.451 ff.).

One particular type of simile compares the fight of the epic heroes with wild animals attacking the cattle. It is often the lion that provides the point of comparison. It is unlikely that there were lions in Greece when the poems were composed, but the lion is a well-known motif in the art of the Greek Bronze Age and reappears at the time of the late geometric style; it is, of course, also well known in the literature of the ancient Near East. As this animal was looked upon as the wildest of them all, there is no reason to wonder why it should play this role in the epic tradition. Dunbabin (1957) has a remark that Homer's lions are never heard to roar, from which he deduces that they were probably literary or iconographical borrowings. Deducing *e silentio* is always dangerous, but his observation is strikingly accurate. It does not follow that the cattle under attack are of foreign origin too. On the contrary, the poet has introduced the wild animal into the reality that he and his audience know. Otherwise the simile would not present the lucid picture designed to illustrate the main trend of the story.

The wild animals will attack a herd of animals and kill a single one, an ox (*Il.* 11.172), a lamb or a kid (*Il.* 16.352 ff.); or they will go in search of stables and paddocks (*Il.* 5.554 ff.; 11.548 ff.) where the cattle are supervised by men and dogs who fight back unless, like the shepherd (*Il.* 5.136 ff.), they are frightened and go into hiding. These similes do not mention villages or any other regular buildings. *Stathmos, aule* or *mesaulos*, in theory, may be found anywhere, as the words do not necessarily signify anything other than pens or paddocks. With this connotation we shall meet them again in the few sources that specifically mention transhumance. In all probability, such were the ornamentations with which Hephaistos embellished the shield of Achilles since *stathmoi*, thatched huts (*klisiai*) and pens (*sekoi*) are found in a lovely valley in the mountains (*en kale besse*). In the same place there is a description of the cattle hurrying to the pasture (*nomonde*) direct from the dung (*apo koprou*) (*Il.* 18.575 ff.). Elsewhere, they hurry home to the dung (*es kopron*) where the calves are awaiting them so eagerly that the pens (*sekoi*) cannot restrain them (*Od.* 10.411). In both cases, presumably, we are dealing with fenced areas which give shelter for the night; in the latter case the calves are kept there whereas the cows graze farther afield.

There is a noteworthy agreement between Aristotle's view of domestic animals as being herd animals, and Homer's similes. Admittedly, the epic genre likes to describe conditions as particularly glamorous, and quite often the animals live under a system of ranching, guarded by dogs and shepherds. This requires very special conditions.

The animals require water, and it is scarcely a coincidence that the shield of Achilles also shows an ambush where the herd of cattle is attacked at a river (*Il.* 18.520 ff.), just as the herd of cattle mentioned above is also attacked at a river where there are rushes growing (575 ff.). With this, we may compare the herd of cattle comprising a multitude of animals living in a water-logged area with perennial growth of grass (*Il.* 15.630 ff.).

Animal husbandry of this kind cannot, of course, be maintained everywhere in Greece. It requires space and reasonable grazing. We have already looked at the landscape of the Plain of Marathon where the river estuary forms a swamp with grazing and water for the cows, whereas sheep and goats are relegated to the less opulent growth on the untilled heights which are not arable (cf. pp. 14 ff.). It is probably not a coincidence when, as mentioned above, Nestor aims his predatory cattle raid at Elis where rain is more plentiful, nor that the famous Epirote cows are to be found in the more humid climate in western Greece. If, in general, we consider references to cattle, we may conclude that Thessaly and the Peloponnese were known for their horses, oxen, sheep and goats whereas Euboea and Boeotia have in fact taken names derived from the word for oxen. As shepherd-country, Arcadia in the central Peloponnese was renowned, and from there the best mules were obtained. Sheep and goats were found everywhere, even in the driest areas like Attica and Megara. Likewise, pigs could be kept everywhere, although to a lesser extent, but we have noted earlier the category into which Aristotle placed them. In greater numbers they could be kept in wooded areas, which were probably not so scarce as was argued in earlier literature on the subject. A precise charting of the distribution of cattle-breeding cannot be attempted here, but it should be noted that the locations mentioned as far as larger cattle are concerned, all benefit from greater precipitation (cf. Figure 1).

This coupling of information derived from Aristotle and Homer may seem bold. To some extent, it does limit the theory proposed by Snodgrass that, from being an agricultural society in the Bronze Age, Greece reverted to rather more pastoral patterns of culture in the so-called 'dark' centuries, then became an agricultural society again in the historical period.[88] The dark centuries are poor in archaeological findings, and the country was perhaps more or less depopulated. However, the life of the shepherd is no more attractive than that of the farmer. There does not seem to be any reason why drastic alterations of exploiting the soil should have taken place, even if untilled ground could be used for areas of grazing where the climate, that is to say the summer precipitation, would warrant it. It is true, of course, that the consumption of meat occurred much more frequently in the Homeric epics, but we prefer to interpret it as an expression of the specific epic glamour that surrounded the lives of the heroes. It was a widespread ancient theory that cattle-breeding constituted a somewhat more primitive stage than agriculture, but this is scarcely so.[89] Rather, it is a question of two different ways of exploiting different types of soil. Extensive cattle-breeding requires a great deal of space, but it should be borne in mind that the Mycenaean tablets, in their turn, testify to the existence of considerable numbers of domestic animals. Here it would seem reasonable to discuss the problem of transhumance.

TRANSHUMANCE

Transhumance is a type of cattle-breeding where cattle change between summer grazing in the mountains where precipitation, and therefore also grass, is plentiful, and in the winter season grazing in the lowland where winter rain allows for growth. Transhumance is known from many places in the world and in many different periods. In the Mediterranean world, we are fairly well acquainted with transhumance in Italy in antiquity and in more recent periods; in Spain in the Middle Ages and from more recent times; and from North Africa, Provence and in Greece where in particular the Sarakatsani and the Vlaches from northern Greece have solicited interest. They live in mountain villages and graze their herds during winter all the way down to the Peloponnese.

In his tragedy, *Oedipus Rex* (ll. 1121 ff.), Sophocles makes two shepherds meet on Mount Kitairon where they have their herds grazing through the summer. During winter they have them grazing near their respective home

---

88 Snodgrass 1980, 35; further 1987, 188 ff.
89 Aristotle, *Politics*, 1256a.

towns, Corinth and Thebes. One shepherd is a slave, the other a hired worker. In other words, they are both of inferior status and tend another man's herd. It is specifically stated that they graze their herds during summer for six months each year. This is what is known as 'transhumance normale'. Its opposite 'transhumance inverse', is hinted at in the seventh speech of Dio Chrysostom where the homestead is in the mountains of Euboea.[90]

Transhumance does not require much in the way of solid buildings. From Sophocles and Dio we find *epaulos*, *aule* and *stathmoi* and perhaps *skene*, a hut. Traces of these more or less perishable pens and primitive buildings are often found in modern Greece, and there is no particular reason to assume that such installations were more elaborate in antiquity.[91] In our own time we know of cases of very brief changes of pastures when transport takes only a few days, and where there is an established connection between winter and summer stations. A vivid description of this, from southern Argolis, has been given by H.A. Koster (1976). The question is not whether transhumance existed in ancient Greece; the question is, exclusively, of its extent and importance. Apart from very few literary descriptions, we have some epigraphical sources which mention the right to use the pastures for summer grazing, *epinomia*. Grazing is restricted to citizens in the territory where the pastures are to be found, but as a privilege the right may be conveyed to strangers. Such areas will often be located in the border-area between two states and may give rise to controversies. An example is found in Thucydides 5.42, where the conflict concerns a pasture common to the Athenians and the Boeotians at Panaktos and in *Hellenica Oxyrhynchia* (8.3), when the controversies between Lokris and Phocaea turn out to be the cause of the Corinthian War in 395/4. Single treaties between states are also handed down epigraphically; by these, the passing of herdsmen with their herds through foreign territory was regulated. So, the boundaries of the city-state were not insurmountable, but this did require authorization on the part of the government of the state.

Whereas, in earlier research, by simple analogy with more recent times, transhumance was used as an explanation of a number of phenomena and taken for granted, we may now observe a greater scepticism. This applies to archaeologists dealing with prehistory[92] as well as to a group of scholars who devote themselves to the historical period in particular.[93] It is probably not possible to determine exactly how great a role this type of cattle-breeding played, nor how many were involved. Everything indicates that the shepherds were of an inferior social status, and that the herds did not belong

---

90 Transhumance is often discussed, see esp. Georgoudi 1974 and the discussion by Hodkinson 1988 and Skydsgaard 1988b.
91 Kouremenos 1985.
92 Cherry 1988.
93 Garnsey 1988b.

to them. In this connection it is better to disregard the Hellenistic bucolic poetry as a source: partly because it pertains to a period later than that with which we here are concerned, and partly because it reflects an idyllic description of the herdsman's life which most likely does not correspond to reality. It is our opinion that larger numbers of animals in the herds were moved to the relatively scarce habitats that would ensure sufficient fodder throughout the year, or else the animals were moved from place to place wherever an adequate amount of fodder was available.

Sheep and goats lamb in early spring. Shearing takes place somewhat later, immediately before the animals are led to pastures on higher land. We have already shown how simple the production of cheese is with regard to tools, and the transport of cheese to places where there is a market offers no difficulty. Thus, there are no technical difficulties connected with utilizing the secondary produce of the animals. It should also be noted that, if the fodder situation allows it, such animals which might be required, for instance for sacrificial purposes, can be retained near the city.

OTHER SOURCES CONCERNING ANIMAL HUSBANDRY

Although we have quite a few references to cattle generously spread throughout our tradition, it is only on rare occasions that we are able to make direct deductions concerning animal husbandry. For instance, we find the following *gnome* by Theognis (ll. 183 ff.): 'We seek noble rams, donkeys and horses as breeders, but a good man does not hesitate to marry a wicked man's wicked daughter . . .' (cf. l. 1112, with no reference to the domestic animals); but apart from revealing a knowledge of deliberate breeding, the quotation tells us little. We must look for more substantial information.

In the Attic forensic speeches we find a couple of specific bits of information. In a litigation concerning a will, it seems that someone has wasted no time in selling out part of the assets, namely, a piece of land, a bath house in the city, a house in the city, and in addition a herd of goats, with a herdsman, two teams of mules, along with all the artisan slaves – all told at a value of more than 3 talents. We cannot tell whether the mules were designed to provide city-transport or work in the countryside, and the number of goats is not indicated specifically, but the price paid for them and for the herdsman amounts to 13 minas (Isaeus 6.33). Furthermore, the same author (11.41) mentions a fortune comprising a piece of land in Eleusis, valued at 2 talents – 60 sheep, 100 goats, tools and a mount which the owner had been using when *phylarchos*, along with other equipment. In a speech by Demosthenes (47.52) it is mentioned that the defendant and his accomplices stole 50 sheep, along with their shepherd and his assistant, as well as some domestic utensils, including a bronze hydria. Not satisfied with this, they trespassed on the owner's piece of land (*chorion*) where he lived, near the hippodrome, and caused several instances of damage. In the latter case we should note that first the sheep were stolen, and then the trespassing on the man's ground took place. Consequently, it would seem that the sheep had been grazing elsewhere.

Scholars interpret these passages very differently. Burford Cooper (1977/8) would prefer to regard the Demosthenic passage as evidence to show that the sheep were grazing 'on public land', whereas Hodkinson would attach importance to the fact that the sheep were grazing near the farm and are to be looked upon as part of it – like the herds mentioned in the two passages from Isaeus. In fact, he speaks about 'agro-pastoral farms'.[94] We intend to stress the point that in these speeches domestic animals are dealt with as separately valued entities which are not sold or evaluated along with other property. This is scarcely a coincidence. In the same way, we find domestic animals evaluated separately on the Attic stelai, and there is nothing to indicate that a country estate is sold with its stock of animals. Thus, Panaitios has 2 draught oxen and 2 unspecified oxen, 4 cows with calves (how many is not known), 67 goats and 84 sheep, both registered with kids and lambs respectively, without any indication of number.[95]

All in all, we must conclude that these sources, few in number, do not by themselves provide a clear picture of the kind of animal husbandry in ancient Greece, but they do allow us to stress the point that, in any case, sheep and goats appear as herds and are treated as separate entities in declarations of property and sale, sometimes together with a herdsman. This cannot be used as evidence of an essential connection between agriculture and animal husbandry, apart from the fact that frequently the land and the cattle belonged to the same owner. As it remains uncertain whether farmers normally lived in farms in the open country or in buildings of rather a more urban character, perhaps in the main city itself, no significant conclusions can be drawn on the basis of this evidence. Nevertheless, we do know that, in time of war, rural districts had to be vacated and the population, and sometimes the animals also, evacuated to the city. People usually chose to send the animals elsewhere, into the mountains or, as the Athenians did during the Archidamian War, to a different state, such as Euboea or some of the islands. Thucydides (2.14.1) mentions sheep and draught-animals (*probata kai hypozygia*), so that we cannot be in doubt as to which animals were at stake. Thanks to Andokides we know that Athens was heavily overcrowded by refugees, sheep, cattle and chariots (fr. 3). This entire complex of problems has been thoroughly dealt with by Hanson (1983) to whose work in general a reference will suffice.

Against this conclusion that, apart from draught-animals, cattle was counted by herds and lived in herds, it may well be argued that only large numbers were sufficient to attract the interest of our sources. This is possible, and we have to admit that the four cows and their calves belonging

---

94 Hodkinson 1988, 38 ff.
95 Meiggs/Lewis 1969, no. 79 B.68–74.

to Panaitios indicate a relatively modest animal husbandry in connection with a country estate. However, such an estate is not mentioned. The two work oxen are kept *en Ar–*, and no satisfactory supplement has yet been offered. Beehives are also mentioned with an indication of their location, but the hives are not necessarily to be looked for at the ground owned by Panaitios. Consequently, we are unable to ascertain whether they have any relation to Panaitios' real property.

Small units of cattle are seldom mentioned. A single case should be mentioned: Aelian quotes one Aescylides, of whom we otherwise know little, as a witness to the effect that people on Keos had but few sheep (*oliga probata*) which they fed by tree-medick (*cytisus*), leaves of fig, leaves of the olive tree, follicles of pulse, etc. These sheep yielded a great deal of milk from which the owners produced an excellent cheese, and this brought them a very advantageous price (*De Natura Animalium* 16.32). Hodkinson sees in this 'an excellent example of labour-intensive, integrated agro-pastoral land-holdings on which fodder crops and the residues of pulse and tree cultivation were utilized for rearing animals'.[96] With this, we cannot agree. Aelian's interests are largely associated with *mirabilia*. Reading the context will demonstrate it clearly. As a cause for the unusual stock of sheep it is adduced that the soil of Keos is poor and without *nomos* where you would have expected to find the sheep grazing. Therefore we must regard this example as an exceptional case which confirms the rule, a *mirabile* on a par with the others which are mentioned. However, this does not exclude the possibility that many – especially people who were less well off – may have had a limited number of small cattle and perhaps some draught-animals. They may have grazed on marginal land where crops could not be raised, and naturally they would have been given what was at hand – for instance pods, leaves, etc. In modern Greece you often see branches pruned from olive-trees collected in heaps, whereupon goats will strip the leaves from them. Once the wood is dry, it will be used as fuel, sometimes in the form of charcoal. A modest herd of sheep or goats will be sufficient to satisfy the need for milk and especially cheese in a smallish household, but it does not seem to warrant a description like 'agro-pastoral farming'.

Likewise, a modest amount of pig-keeping may be based on such fodder as kitchen refuse. One might recall the passage in the *Acharnians* by Aristophanes where the farmer from Megara wishes to sell his daughters by pretending they are pigs (ll. 736 ff.). Even in Athens you might encounter pigs, if indeed we can trust Plutarch's anecdote about Socrates meeting a herd of pigs in the street (*De Genio Socratis* 580 E). If we wish to find a modern parallel to animal husbandry of this nature, it would be natural to visit Methana where husbandry is described as follows:

Besides growing a wide range of crops each household owns a few

---

96 Hodkinson 1988, 46.

sheep and house goats for milk, meat and wool and hair, and a donkey and/or mule or two for draft and transport. In the old days nearly every household had a pig. Now there are only two families with brood sows. Also in the past, numbers of sheep and house goats were larger and several families had flocks of *ghidia* (range goats).

(Forbes 1976, 239)

It should be remembered that Methana is one of the Greek districts with the lowest precipitation. In other areas where precipitation is more plentiful the possibilities for maintaining larger stocks of cattle are much more favourable. It is, however, significant that mules still had to be imported to Methana at the time when the investigation was performed since no horse existed!

TRANSPORT

A comparison with modern conditions forces us to consider that draught-animals, too, had to be bred for transport that need not necessarily have to do with agriculture. In larger cities it must have been essential that donkeys and mules should be available for internal transport by chariot, and overland transport usually took place by mule. Similarly, heavy transport, especially of building material, was made possible by means of teams of oxen. We cannot possibly estimate the number of animals thus employed, but it must have been considerable. Osborne (1987, 14) points out that there must exist a close connection between 'the agricultural year' and 'the construction year', – transport of building material takes place during the months when the oxen are not occupied by ploughing. The empirical material is not very adequate in favour of 'the construction year', consisting of a few inscriptions indicating expenses connected with transport, mostly from Eleusis. It would be strange if plough-oxen had not been used when available. To our knowledge, few attempts have been made to estimate the quantity of stones used for the construction of Greek monumental buildings. One attempt at an estimate arrives at a figure of more than 20,000 tons used for the construction of the Parthenon.[97] For this, many teams of oxen must have been required, in whichever way blocks and drums may have been transported. It bears comparison with the Attic honorary decree which was raised in the year 329, commemorating Eudemos of Plataea. Among other things, he contributed 1000 teams of oxen for the work at the stadion and the Panathenaic theatre.[98] This gave to him and to his descendants the title of *euergetes* as well as the right to own land in Attica, and also the right to perform active service in case of war, and to pay *eisphora*, in other words all civil rights except the right to vote and eligibility.

97 Stanier 1953.
98 Tod 1948, no. 198.

As the inscription has been interpreted in different ways, it calls for a closer analysis. We agree with Burford (1960) that Eudemos of Plataea is not likely to have been the owner of 1000 teams of oxen. Rather, he places them at the disposal of the construction work and may have hired them from perhaps as many farmers. Thus, a private citizen has taken upon himself the entire organization of a vast contractor's enterprise, probably as a voluntary donation. Whether these were teams of oxen together with chariots, is probably more than doubtful ('carts and pairs of oxen', as Tod (1948) maintains). Burford points to several sources that show that several teams of oxen were necessary in order to haul a single load. Furthermore, she suggests that, perhaps, Eudemos did not provide the oxen but put up money to hire them; she suggests as a possibility that we are not, in fact, dealing with teams of oxen in the proper sense of the term, but that we have to envisage 1000 'yoke days'. Both suggestions seem problematic. Apart from the oxen, Eudemos provided 4000 drachmas to the war (*eis ton polemon*). Why shouldn't the decree have continued to state that he had placed an amount to be used for the transport if that was meant? As far as 'yoke days' are concerned, one would have liked to see just a couple of examples to show that *zeugos* could have this connotation, but the Eleusis accounts register payment per team per day over a specific number of days. It may be difficult to visualize the actual situation, but the oxen would not necessarily have to be present simultaneously. It is questionable whether a team could manage heavy transport over a long period day after day, but we do not know the nature or the duration of the work. It may have been a matter of removing earth and other material from the excavations while the theatre and the stadion were being erected, or it may have been transport of building material for the building operations.

One should remember that the draught-animals require fodder as well as water. Inasmuch as they are out of their agricultural context, then fodder has to be transported to the resting places or at least left in caches. The responsibility for this part of the job rested on people of whom we know nothing, but it must have called for considerable organization which should not be underrated. The purchasing of fodder would probably have had to take place, whereby the transaction becomes even more complicated. It is also possible that the individual teams carried their own fodder, sufficient for at least a few days, but the oxen require an ample amount of fodder, provided we may rely on Cato's rations (*De Agricultura*, 60).

All in all, Osborne seems vaguely optimistic when, as it seems, he reads between the lines that transport of building material is simple because the oxen are nearby. No major undertaking is simple in a pre-industrial society, but looking at the results, we have to conclude that this aspect of the matter was also solved. It is only when trying to visualize the situation that we see the difficulties involved. This is probably what Plutarch did when he wrote the famous chapter (12) in his Life of Pericles where he enumerates the numerous different artisans employed in the Periclean building programme. Naturally, transport tradesmen are also mentioned, but the chapter is a reflection rather than a rendering of an earlier source. It should be remembered that the building programme included the Long Walls, the erection of which was probably initiated before 460. Here, too, heavy transport was needed for building material as well as rubbish to be used as foundation in a swampy area (cf. Plutarch, *Cimon* 13). Rebuilding these walls also called for transport as mentioned in inscriptions which, for example, mention payment for teams of oxen (Tod 1948, no. 107). The erection of the Long Walls served to protect overland transport from Piraeus to Athens, and transport along this line must certainly be called heavy, especially when imported grain was in question.[99]

It is a dogma often repeated that heavy transport over land was an economic impossibility in antiquity. Nevertheless, this expense was met with when needed: modern historians, we feel, have often let themselves be blinded by a contemporary cost-benefit evaluation which was far from the ancients. Burford's thorough analysis of the sources shows it, and we must bear in mind that, naturally, our sources are mainly concerned with building activities because these were public and often connected with the erection of sacred buildings.

We have no way of guessing at the number of oxen which were employed for tasks outside agricultural service, nor can we estimate the total number of draught-animals. We can only surmise that procuring such animals at certain fixed points of time, in great quantities, as well as keeping them fed during transport, must have been extremely complex questions of organization; however, such problems are far beyond the spheres of interest in the sources which are at our disposal.

Light transport, with donkeys and mules, is generally speaking far less known. An amusing, albeit late source, is pseudo-Lucian's novel about Lukios who is transformed into a donkey and changes owner incessantly, therefore being commanded to perform the most diversified types of labour. Apart from the fact that, in our period, it appears that donkeys have not been made use of in working the grain quern,[100] all other work can easily be referred to our period. In particular, we may think of the transport of fuel, mostly wood and charcoal, particularly in connection with mining,[101] but donkeys and mules were probably met with everywhere. Once again, we may refer to the vase painting in the Louvre (Plate 3.3) where, beside ploughing, we see a two-wheeled cart loaded with two large vessels and drawn by

---

99 In his speech against the corn-dealers Lysias mentions that they were not allowed to buy more than 50 *phormoi*. Unfortunately we do not know the measure, and it would be rather bold to assume that the *phormoi* were the baskets loaded upon a donkey or a mule.
100 The donkey-mill is probably a Hellenistic invention, Moritz 1958.
101 Phainippos is said to possess six donkeys transporting wood and giving the owner more than 12 drachmas every day (Demosthenes 42.7). This is considered an exaggeration by de Ste Croix 1966. For fuel to the manufacture of pottery see Hannestad 1988.

two mules. Attention should be paid to the donkey which is moving in the direction of one handle. These hardy animals require little care. On the island of Lesbos Skydsgaard has observed mules that move freely in the vast olive groves throughout the summer. They seem to manage by gnawing grass between the trees and will go in search of human dwellings only when they need to drink. At the time of olive-harvesting in winter, they find their proper task, that is, they are needed to bring back the picked olives, often from trackless sites. Therefore, if there is a problem concerning these animals, it does not seem to be feeding them, as for the greater part of the year they shift for themselves; rather it is the expense of acquiring them. The maintenance of a reasonably large stock of mules to undertake any odd job must have required a very deliberate programme of breeding elsewhere.

It is no wonder that the trade in animals, particularly draught-animals, as presupposed here, is seldom mentioned in our sources. In a brief article Grassl (1985a) has collected the most important references to sources concerning cattle-trading, and, as was to be expected, by far the largest number concerns trade in sacrificial animals. These are also the main subject of Jameson's contribution to the Congress in Bern (Jameson 1988): it is especially the large cultic centres that have need of a multitude of cattle of a high quality, but naturally local cults have also required animals.

The epigraphic material from Delos shows that here there was an unusually large amount of cattle-keeping. This must be ascribed to the specific ready market for the shrine. Since this is a small island community, cattle-keeping becomes much more apparent. Owing to the much easier transport of animals over land, it has not attracted the same attention in other sanctuaries. The question of the gods and their cattle will be discussed later. The less conspicuous trade in smaller draught-animals has been of less interest in the sources, a phenomenon that shares the vicissitudes of fortune which apply to so many everyday events in antiquity.

6

AGRARIAN SYSTEMS

In the present analysis of the sources available our interpretation has been a continuation of earlier research as represented by, e.g., Jardé (1925), Semple (1932), Michell (1940) and Hopper (1979). The sparse contemporary sources mention the biennial system alternating between fallow and crop; systematical growing of specialized fodder plants does not appear to have been commonly practised, and consequently animal farming on a greater scale has been limited to special ecological niches. More generally speaking, Eric Wolf (1966) has contrasted 'mixed farming' or 'balanced livestock and cropraising' with the so-called 'Mediterranean ecotype', which fits fairly well with the picture to which our analysis leads us.

Against this, an alternative model has now been set up, most recently adopted by Peter Garnsey in *Famine and Food Supply in the Greek and Roman World* (1988a), with a large section dealing with Attica: 'A rival picture of ancient Greek farming is now gaining ground among historians and archaeologists according to which small-scale intensive mixed farming was the norm in densely populated Attica in the Classical period' (93). Although, in the preceding chapters, we have occasionally felt compelled to argue against this so-called new or alternative model, it will be necessary, as the conclusion of the first part of the present study, to analyse this with a view to establish a complete survey.

The starting point for the alternative model was P. Halstead's essay, 'Counting sheep in neolithic and bronze age Greece' (1981), where he presents an interpretation of the earliest phase of agriculture from the Neolithic Age to the end of the Bronze Age. Neolithic agriculture is summarized on page 335:

> Small-scale stable gardening with crop rotation and regular manuring. Pulses seem to be as important as cereals in the agriculture of prehistoric Greece and, arguably, this departure from the traditional picture is only practicable under a small-scale regime. Similarly it is argued that archaeologically observed regularities in the proportions of different livestock species at neolithic settlements reflect the integration of

animal husbandry into just such a pattern of land use.

The prerequisite for this archaeological interpretation of archaeological data is the relatively widespread, and not very numerous, population which is characteristic of the earliest phase when it was not yet necessary to include marginal land:

> It is suggested that community size is of more direct relevance to patterns of land use and that the growth of larger communities in the later Neolithic and Early Bronze age may have undermined the earlier gardening economy and so have favoured innovations such as the growing of olives (especially in southern Greece and Crete) and ploughing (perhaps particularly in northern Greece).
>
> (p. 335)

Thus we find that the new model springs from an archaeological interpretation of Neolithic agriculture with its sparse and more widespread substratum of population.

The next step is taken by T.W. Gallant in his 'Agricultural systems, land tenure and the reforms of Solon' (1982). Here the agricultural system is discussed, with an inspiration borrowed from Esther Boserup (1965) but, as it seems, with a very unfortunate abridgment of the epoch-making study of this scholar. Gallant writes:

> (1) Short fallow: the period of cultivation is a year or two as is the period of fallow. Only grasses have a chance to grow on the fallow land. This system requires the plough to break up the grass cover, and manure to ensure a steady and high level of production.
>
> (2) Annual cropping: the same plot of land will have two crops planted on it per year, usually a fodder crop then a cereal crop. There will, however, be a period, about a month or two, when the land is not cultivated. This system requires intensive labour in relation to output, multiple ploughing, and manure.
>
> (3) Multi-cropping: the same plot of land bears two or more successive crops, and the period of fallow is negligible. This system is very labour-intensive and in most cases requires irrigation.
>
> (p. 113)

First, we have to emphasize that Boserup's models are almost exclusively based on empirical data from tropical areas and not from the Mediterranean climate, let alone from Europe north of the Alps; but this is of minor consequence. To us, the decisive factor is that, as shown previously (see above, note 13), Gallant misinterprets Hesiod, *Works and Days*, 606–7, thus introducing a crop of fodder plants not contained in Hesiod's text. It is characteristic that, in order to comment upon Hesiod, he is compelled to introduce Columella, Pliny and Palladius, a method which, in our opinion, is inapplicable. As far as we can see, there is nothing to suggest that Hesiod knows of plants grown exclusively for fodder. For this reason we shall refrain from discussing his interpretation as we feel that it is based on substantial misinterpretations of the literary sources.

Jameson has also been inspired by Boserup in his pioneering article, 'Agriculture and slavery in Classical Athens' (1977/8). An increase in population can lead to colonization, but as Attica does not appear to have participated in the great race for colonies, what remains is to utilize the natural resources of the country; as for agriculture, this means intensification, diversification and specialization. Marginal land is taken into use, cultivation of vine and olive is introduced, and several crops are raised on the same pieces of land. Thereby the yield grows, but it requires an increase of labour. Here, slavery enters the picture as a possibility. Agricultural slaves, as we know, are seldom mentioned in literature, but, as Jameson has elegantly suggested, the reason why they are mentioned so rarely is that they were to be found everywhere. Of course there were other ways by which labour could be increased than by means of slaves, but we should remember that the slave has the advantage of not having to support a family; therefore, the yield should be shared with none other than the owner. Not surprisingly, Jameson's argument has not remained uncontradicted, but this is a matter of interpretation, not a matter of including new sources.[102]

For our purpose it is decisive that Jameson questions the alternating fallow and crop, the biennial system. He quotes a number of leasing inscriptions which indicate that some have fallen for the temptation to sow crops each year on leased soil; but it is rather a large assumption to deduce, from this, that this was a common phenomenon. We must weigh the reliability of the source attributed to the literary tradition which sustains fallowing every other year against the single cases which may be adduced to the contrary. We have previously mentioned our doubts over the growing of pulse being regarded as so important, and we have emphasized the inadequate knowledge of Theophrastus when dealing with the ability of pulses to collect and store the nitrogen of the air. He knows that some pulses are useful, but not why (*Historia Plantarum* 8.9.1), and when it is argued that in Macedonia and Thessaly the entire plant is ploughed into the earth when in bloom, it is clear that the process is still at the experimental stage (how you can plough down plants with an *ard*, is a problem in itself).

S. Hodkinson (1988) attacks the problems from the point of view of animal husbandry. We have taken the opportunity, particularly in the previous chapter, to emphasize how we disagree with his views, especially with his notion of the 'agro-pastoral farm'. The circumstance that the same owner entertains animal husbandry as well as agriculture does not mean that these two branches are integrated in some sort of symbiosis corresponding to Eric Wolf's 'balanced livestock and cropraising'. Hodkinson's paper did

---

102 Cf. Wood 1983 and 1988. Jameson has answered the critics and elaborated his views in Wells, ed., 1992.

not remain unopposed at the Congress at which he presented it, but it is of course difficult to adduce a decisive proof in favour of a point of view based upon one interpretation of the sources. The decisive argument against the alternative model is its assumption that Greek agriculture was in a position to set aside essential areas for the cultivation of fodder plants. As is well known, man and domestic animals are competitors with regard to food,[103] and animal production in Greece depends on crops that could otherwise serve as food for humans. It seems illogical to stress 'the peasant character' in Greek agriculture while at the same time you introduce a stock of animals to be fed on the crops from the same very limited land. True, there are indications that some operated a slightly different type of agriculture than that traditionally applied, but to what extent this tendency may be said to have been generally applied, remains an open question. Skydsgaard (1988b) adduces a number of reasons for doubt, but we shall not quote them all. If, for instance, the question concerns the growing of alfalfa, it is clear that the plant was found from the fifth century onwards, presumably imported from Persia, but how widely distributed was it? Theophrastus mentions it but seems not to be too familiar with it, and Aristotle has a direct warning against employing alfalfa as fodder for milk-giving animals, as it will stop the flow of milk (*Historia Animalium* 522b). This surprising piece of information in no way corresponds to modern experience; the warning probably survived only because alfalfa was not a common crop. Probably lucerne was reserved for the horses. The average peasant is not likely to have been able to spare a major area for a fodder plant over a number of years and alfalfa is a perennial crop.

An important factor, however, is this: a fairly large stock of domestic animals yields an increased production of animal manure from which the cultivated areas will derive a direct benefit, and increase vegetable production. This requires the presence of animals fed in stables so that their manure can be gathered and spread in sufficient quantity. Xenophon's dungheap, essentially, seems to be a compost heap of vegetable refuse (*Oeconomicus* 20.10 ff.), and the famous dungheap, where Odysseus finds his faithful dog Argos, consists of manure from draught-animals such as mules and oxen (*Od.* 17.297). If the herds of cattle are centred round the ecological niches away from the arable land, regardless of whether there are wetlands for the oxen or maquis for the goats and sheep, there will be precious little manure to spread. In order to explain the greater concentration of potsherds and other artefacts in the vicinity of built-up areas, archaeologists have suggested that such remains could well have been distributed along with manure from the stables.[104] This is of course a possibility, but ploughing over thousands of years could presumably have contributed to a further spreading.

103 Jongman 1988.
104 Wilkinson 1982, 1989; Snodgrass 1991.

One must now ask whether it is possible to uphold two positions, one against the other, which would entail two different interpretations of Greek agricultural economy. In this connection it would be sensible to refer to Boserup (1965); in chapter 6 of her book she deals with 'The coexistence of cultivation systems'. Cultivation systems should be conceived of as models, not as stages in an evolution which, recurring at fixed times, lead towards an imaginary goal. It may be important to note that, in Greece, it is only during the most recent periods that a transition from 'dry' to 'irrigated' culture has taken place, that is, to a culture where water can be found and distributed by means of drilling and the use of petrol pumps. In this way a long and tortuous path has been covered, but artificial irrigation is, of course, not applied everywhere. This phase presumably had its beginnings in antiquity. Homer has a simile of the man who, with his pickaxe, conducts water from a spring to trees and gardens (*Il.* 21.257 ff.); and occasionally, in archaeological literature, we find references to structures that may have some connection with irrigation, but they do not seem to have been in common use. Greek water-courses tend to dwindle in the hot summer when artificial irrigation is needed; only later were means found by which water could be raised artificially, such as the Archimedean screw and the water-raising wheel, which seem to have been in use in quite different areas. Greek agriculture was essentially the dry-field system.

The scattered suggestions of artificial irrigation will not allow us to deduce that Boserup's model 'irrigated agriculture' had been arrived at, nor must occasional references to pulse crops, and so on, which were used as fodder for domestic animals, lead us to believe that the biennial fallow, so well attested, had been abandoned in favour of a more systematic rotation of crops as used by the Romans. With the Greeks this was on a rather more experimental level, often undoubtedly so that land which was closest was cultivated most intensively; for one reason this land was, as a rule, likely to be the most fertile part of the arable soil. It is also the part of the land that received most of the manure or compost that could be produced by a system of agriculture poor in cattle.

According to Boserup, an increase of population leads to intensified cultivation with a larger effort in terms of labour, but it does not seem reasonable to assume that this is based on an extended stock of domestic animals with further competition concerning food. On the contrary, we may ask whether the costly draught oxen were replaced by human labour so that, along with ploughing, an extensive system of agriculture with pickaxes developed, without the use of animal labour. Jameson (1977/8) adduces several examples to demonstrate this (from, for example, New Comedy).

The solution is, of course, that there was not just one, but several different agricultural systems, depending on local conditions. The individual farmer has at all times known very well how his different plots of land could be utilized optimally. Garnsey introduced the new model with these words: 'It is pertinent to ask in respect of subsistence or nearly subsistence farmers, whether they could afford to cultivate only one half of their meagre plots each year' (1988a, 94). We would be quite as justified in asking whether, faced with an increase of population, an agrarian population would be prepared to share the crops from meagre plots with an increased and less economical production of domestic animals, where calories demand much greater resources than in vegetable production. Undoubtedly, by far the greater part of the pulse crops was intended for human consumption, and only whatever may have remained was used as fodder for the few sheep and goats which could be kept close to the cultivated areas. From this concept of agriculture it must follow that we must refrain from participating in the guessing-game concerning normal production. Jameson sets up a cautiously worded calculation based on modern average figures, taken over from Keith Hopkins. Garnsey criticizes earlier calculations by Barbagallo and Jardé and presents new suggestions. In both cases we are dealing with calculations that cannot be substantiated by ancient sources. Not knowing the cultivated area and unable to verify the existence of one and only one system of cultivation, and furthermore not knowing the yield nor the amount of sowing per area-unit, we must conclude that such calculations should be relegated to scholars' desks as some kind of mental exercise. The figures may have an authoritative appearance of likelihood, but more often they are misleading and intended to exemplify a line of argument only. It is a matter of regret that in future textbooks they will be found quoted out of context and removed from their theoretical starting-points as if they were proper figures based on studies of the sources.

The originator of the New Model, Paul Halstead, resumes the discussion of the problems in an article from 1987; there, he is much more cautious than Garnsey and Hodkinson. The article is extremely thorough but does not contain one single discussion of one single statement from ancient sources. Actually, what we find is rather a series of questions and answers to questions, and as far as we can see, no real distinction between Greek and Roman agriculture is made. On the other hand, there is an excellent discussion of a long series of problems, and it is emphasized that no agricultural structure is unchangeable, not even in a traditional non-industrialized type of agriculture. In general, caution is advocated against direct conclusions from the less developed type of agriculture of our time to that of ancient times, and a very thorough discussion of a series of problems is offered, with ample references to recent research; but as far as we can see, no new light is thrown on Greek agriculture. A number of elements of uncertainty are stressed, uncertainties in which we readily concur. We cannot, however, arrive at a point that would bring us closer to endorsing the new model, which Hodkinson and Garnsey wish to introduce into historical time.

But the new model may also turn out to be a boomerang: if ancient Greece was a series of communities made up of peasants, they would, in order to sustain themselves, have abandoned the so-called 'traditional' kind of agriculture and adopted other and more sophisticated methods, primarily in an attempt to co-ordinate tilling the soil and breeding cattle. Inasmuch as this cannot be seen unambiguously from the contemporary ancient sources, however sparse they are, it can be concluded that at any rate a number of those Greek city-states were *not* primarily populated by peasants. Here, cities like Athens and Corinth are obvious examples. In these cities there must have been ways to make a living other than agriculture. The peasant model may apply to the greater part of the small Greek city-states which rarely appear in traditional history, but it is incompatible with our knowledge of the larger cities. At this point, Garnsey supplies his own ammunition against the Weber–Finley model upon which he so willingly wishes to build. Hodkinson emphasizes that his use of the models,

> of the second model in particular is as a heuristic device for structuring my discussion and for directing attention to neglected aspects of agriculture and pastoralism, not as an *a priori* statement of opinion. A model is a simplified structuring of reality which presents supposedly significant relationships in a generalized form.
>
> (1988, 69, note 2)

As long as the alternative model is allowed to remain on the model stage, it is inoffensive, but when it is let loose on actual history, whatever this may be – as by Garnsey – caution is called for.

Whereas, so far, we have investigated Greek agriculture essentially from technical and agricultural aspects, we shall now have to widen the view and consider it in its relation to the surrounding world and in its interplay with a number of factors created by the community in the city-state.

Part II

STATE AND AGRICULTURE

INTRODUCTION

In the preceding chapters we studied the conditions offered by nature towards agriculture and animal husbandry in Ancient Greece and, on the basis of this, discussed what was produced, and by means of which techniques.

Parts II and III will deal with the question of how, and under which conditions, agriculture was organized and governed by laws and decrees in the Greek city-states. We shall also examine the interplay between Greek religion and agriculture.

It would have been logical to open the discussion by dealing with the relations between gods and agriculture, in the first place because, so to speak, the gods made their appearance before the city-state, and second because the pantheon was common to all Greeks, while city-states varied from place to place. But in this case logic is not applicable because no written source that could enable us to undertake a closer study of the relations between agriculture and the gods exists before the appearance of the city-state. Therefore, it is only in a few limited fields that the latter may be discussed. The consequences of this will be made the subject of a closer discussion in our chapters on gods and agriculture.

The evidence lays down other limitations. It is common practice to introduce discussions of general topics with a regretful statement to the effect that the source material is sparse and centred on Athens which, in most respects, is atypical. With regard to our topics in the following chapters, the matter may be turned round in such a way that we may call the source material comprehensive and particularly rich as far as Athens is concerned. Speeches written for use in court are a special Athenian phenomenon. They provide us with an insight both into laws and norms in general and into the conflict that triggered the particular lawsuit at any given time. Without the conflict no one would, at the time, have committed the topic to writing. An entire series of such speeches have been written in matters concerning maritime trade, and have supplied us with information on the methods by which the Athenians regulated import and export of foodstuffs.[105] We know of rules governing water-courses and roads through a speech addressed to the court by a farmer, who has been sued for having led a water-course away from his land and on to the road, resulting in the flooding of the lands of his opposite neighbour (Demosthenes 55). And a speech which deals with the stump of a sacred olive (Lysias 7) provides us with the rules applying to those who had one of the sacred olive trees of Athene growing in their field.

Speeches of inheritance, primarily those by Isaeus, contain information about Athenian rights of inheritance and about Athenian property in land and cattle. In short, many such speeches have a direct bearing on topics relevant to agriculture; yet many more contain information on similar questions given in passing. It is a coincidence if in such lawsuits one derives an insight into conditions outside Attica.

Part of the evidence, mostly from the fourth century, albeit written in Athens, deals with other parts of the Greek world as well. Xenophon is probably the author among the Athenians who contributes most comprehensively and personally to our knowledge of the part played by agriculture in the Greek world. His *Oeconomicus* is the only one among the writings preserved from his hand which has agriculture as its central topic. As we have seen in the first part of this study, Xenophon gives us a certain insight into agricultural technique, but in particular he shows how agriculture may ideally be organized so as to fit most favourably into the city-state. *Oeconomicus* concerns ideals, but at a level where connection with realities is constantly felt, often with specific reference to Attic conditions. Xenophon's many descriptions of camp-life in his *Anabasis*, *Hellenica* and *Cyropaedia* take us far afield, also geographically. Just as in *Oeconomicus*, they are based on his own experience, and his writings furnish us with welcome information, for instance with regard to the problems of supply for the armies, and how they dealt with crops wherever their campaigns took them. Finally, it is to Xenophon that we owe most of our knowledge about the interplay between agriculture and religion, as far as our literary sources go.

Two main sources on the distribution of land in Greece are the political theorists, Plato with his *Republic* and, later, *The Laws*, and Aristotle, his younger contemporary, especially with his *Politics*. This may seem strange since it was on the theoretical level that these two philosophers laboured with an attempt at constructing the ideal state; but in order to arrive at arguments supporting the ideal state they were bound to observe the states which were actually in existence. They do this implicitly and, fortunately, also explicitly, partly in the form of direct accounts, and partly by way of scattered remarks about conditions in Athens as well as in other city-states. The ideal states themselves are not the subject of this book. If occasionally we touch upon the subject, it will be merely in order to reach a better

---

105 Demosthenes 32–5; 56, with commentaries in Isager/Hansen 1975.

INTRODUCTION

understanding of the city-states that did in fact exist at the time.

It is also from learned circles in Athens, perhaps from the school of Aristotle, that we have the second book of the *Oeconomica*. This work, with its examples to show how, for instance, Greek city-states have managed to extricate themselves from sudden shortages of supplies or other such deficiencies, gives us a close insight into custom as well as conditions for uncommon solutions whenever it was a question of supplying a city-state with foodstuffs.

As for the question of the relation between agriculture on one hand and state and gods on the other, the epigraphical material plays a decisive role. In particular, we find normative texts like decrees concerning, for instance, conditions for the leasing of land or collection of dues, but the inscriptions may also be proper leasing lists or accounts to show income derived from land belonging to the gods, and how the income from such land has been administered. The material is rich and shows a very considerable geographical spreading. As a rule, such material has been found in connection with an archaeological excavation which, in most ways, is a distinct advantage. Where and when these texts are made available to the public does, however, depend on who was in charge of the excavation. Apart from such difficulties, we may state that in general the following applies to the epigraphical and literary material: Athenian evidence lends itself most favourably for interpretation because our background knowledge about that particular city-state is comprehensive by comparison with others. It is felt strongly when we attempt to interpret a unique, long and detailed inscription like 'The Queen among Inscriptions', the law-code from Gortyn on Crete, dated to the fifth century. In contrast to what was the case in Part I, archaeological material, apart from inscriptions, plays a remarkably small part in this second Part.

Corresponding to the rather abundant evidence, there is no shortage of modern treatments on most individual topics, as will become evident from the following chapters. To our knowledge, no one has attempted to undertake a consistent investigation of these topics from just one point of view, concentrating upon Greek agriculture in antiquity.

7

PRIVATE LAND

In his eminent work on Athenian democracy in the fourth century, Mogens Herman Hansen devotes his first chapter to a definition of the *polis* and raises the question to what extent the concept of *polis* may be understood as the equivalent of our notion of state. His aim is to persuade the reader that any idea of a territorial state should be abandoned. The *polis* was constituted by the community of the citizens, the *politai*, concerning the constitution, the *politeia*, as it was phrased by Aristotle.[106]

But in a book on Greek agriculture it will be necessary to stress that aspect of *polis* which, in his context, Hansen rightly plays down, namely, that the same citizens share a well-defined territory,[107] and that, from the point of view of a Greek city-state, having the boundary-stones removed from the ground would be tantamount to obliteration.[108] When Athenian recruits submitted their oath of allegiance they called upon a series of known deities as their witnesses, but at the end also 'the boundary-stones of their native country, wheat, barley, vines, olive trees and fig trees'.[109] The lawgivers of each city-state did in fact lay down rules as to how the distribution and exploitation were to be regulated. As we shall see, solutions varied markedly from one city-state to the other, depending on the type of government, resources and the degree of independence.

Unlike the question of what techniques were applied in agriculture, topics of a somewhat more political nature, such as the distribution of land and the right of property, have been the subjects of vivid interest among scholars at least during the past century. We shall not pursue the history of research during the entire period, but point at two landmarks, as it were, indelibly planted at the beginning and the end of this timespan respectively, namely, Paul Guiraud and Moses I. Finley.

In 1893 the former published *La Propriété foncière en Grèce jusqu'à la conquête romaine*, with which he won a prize. It turned out to be basic for

---

106 Hansen 1977–81, i, 13–18; 1989a, 20; 1991, 58–64; cf. Aristotle, *Politics* 1276b 2.
107 About the importance of the territorial frontiers cf. Sartre 1979.
108 For Corinth, see Xenophon, *Hellenica* 4.4.6.
109 Tod 1948, no. 204.

all later studies of the subject, regardless of the concept of historical theories adopted by any particular author; so it is for this present book.

Whereas Guiraud laid a foundation, Moses Finley has founded a school. After the appearance of his epoch-making dissertation, *Studies in Land and Credit in Ancient Athens 500–200 BC: The Horos-Inscriptions* (1952), Finley became a focus for attention, someone who was to inspire a number of scholars occupied with the study of land. This became apparent, for instance, at the conference called 'Colloque sur l'oikos' held in Paris in 1967, a conference which by all accounts must have been successful, although apparently no publication containing a full account of all papers has appeared.[110] An international colloquium, arranged by the Centre de Recherches Comparées sur les Sociétés Anciennes in the year 1969, resulted in the publication of *Problèmes de la terre en Grèce ancienne* (1973b), edited by Moses Finley. For most aspects of our topic, this publication is still of vital importance.

While Guiraud is a solid base and requires no specific treatment in our text, Moses Finley can sometimes leave you with a sleepless night. He represented his theses in such a manner that they have almost acquired the character of provocative dogmas which you cannot merely disregard even if you might be of the opposite opinion, or indeed just feel inclined to modify them.

THE STATUS OF LAND

Not all land within the territory of a Classical Greek city-state had the same status. We may find land designated as

1. common (*koine*)
2. public or belonging to the state (*demosia*, sometimes synonymous with *koine*)
3. sacred (*hiera*)
4. private (*idia*)

Although we can guess at the approximate connotation of the various designations, it is not immediately evident how and why the distribution into categories has taken place. In order to examine this question it would be natural to concentrate on cases where Greeks had put their minds to establishing an entirely new city-state, either in the visible world or in the world of thoughts. Also, it would be the ideal procedure to start with the numerous Greek city-states founded in the Archaic period during the first great wave of colonization in the eighth–sixth centuries.

---

110 Finley's contribution was printed for the first time in *Eirene* (1968b).

Unfortunately, this cannot be done because we lack contemporary written evidence.

Instead, we shall start with the ideal state of Aristotle, a relatively down-to-earth theorist in matters of state. In his view, as in that of other philosophers, a division of land into main categories is of vital importance and is determined by the function that those who were to live off the land would have in the imaginary state. In his ideal state he divides the land into private land and common land. The major category, private land, would be distributed among the citizens and safeguard the individual household. Each citizen would have his land divided with one lot near the city, the centre of the city-state, and another somewhere in the periphery at the borderline of the territory. The purpose of the bipartition is to ensure that the will to defend the territory should be equally strong for all the citizens. The common land is divided into the main categories, public and sacred land, whereby we have encountered all the categories known from the Classical city-states. According to Aristotle, public land is to provide for the common messes, the *syssitia* (see p. 132), and he attaches importance to the fact that all citizens should be able to participate. The sacred land is there to ensure that income is available for maintenance of the cult. The citizens of the ideal state would have no time to till the soil themselves, that would be looked after by dependent labour (*Politics* 1329b–1330a).

In the *Politics*, Aristotle also refers to the main features of a proposal for an ideal state set forth a hundred years before by Hippodamos, whom we know better as the famous town-planner (*Pol.* 1267b-1268b). Hippodamos also distinguished between three main categories of land. The sacred ground was reserved for the same purpose as proposed by Aristotle, and likewise, private and public land together were to secure the sustenance of the citizens. But according to Hippodamos such citizens as owned land were to sustain themselves entirely from their private land, whereas public land would supply a different group of citizens, namely the soldiers. Aristotle regarded the Hippodamian system as doomed for reasons that are not our concern in this study.

By necessity, according to Aristotle as well as to Hippodamos, public land would have to be cultivated profitably. It therefore stands to reason that it was not just a matter of surplus land. The concept of surplus land does not belong at all in an ideal state where the philosophers disposed of all land at once.

In the colonies, as they existed in reality, there is little doubt that surplus land was to be found, once land had been set aside for the individual use of the colonizers and for the gods. So we assume that private as well as sacred land had similar functions here and in the ideal states, whereas the ideal states differed from one another as well as from the major part of the Greek city-states in the way public land was put to use.¹¹¹ Admittedly, a phrase of Aristotle's, slightly cryptic to our ears, may be interpreted to suggest that in his time, on Crete, a system operated with use of public land as visualized in his ideal state.¹¹² A Gortynian inscription from the fifth century (*IC* 42b) testifies to the fact that a section of public land has been made available for planting (*phyteusis*); however, the farmers are not allowed to sell or pawn it.

As a rule the public land – the surplus land – was land that the community could dispose of; part of it could, for example, be assigned by voting to the category of private or sacred land. In general terms, we know of this from pseudo-Aristotle, *Rhetorica ad Alexandrum* 1425, where advice is offered as to which arguments are particularly well suited to sway the opinion of an assembly when state finances are concerned. Before direct taxation is imposed, it should be investigated whether any public areas are left unexploited, that is if there are public areas that neither yield any profit, nor have been set aside for the gods.¹¹³ In that case, one might establish an income either by selling them or by leasing them out. This type of income is characterized as 'the most common', since it derives from sources belonging to the community as a whole.

When we speak of arable land it is, therefore, not unexpected to encounter one set of regulations that concern private land, and another that apply to sacred land, but no set of rules for the category of 'public land' in its entirety. Consequently public land will not be dealt with as a separate category here. The subject of this chapter will be the category of private land which was a special concern of Greek legislation. The land of the gods will be dealt with in Part III.

THE COLONY

In the ideal state envisaged by Aristotle, all citizens, as mentioned above, were to start with equal lots of private land, and for reasons of defence each was to have a set of lots, one placed close to the city and one in the periphery. The principle of equal distribution of land to colonists was well known and undoubtedly characteristic of Greek colonization.¹¹⁴ We shall look more closely at the only example at our disposal, where rules governing the distribution of land in a new colony are known in detail. We shall then approach Athens, a city which in the opinion of the citizens had always been

---

111 In the first of the ideal states of Plato there was no privately owned land. The citizens all got their support from the public land, Plato, *Republic* 416d.
112 Aristotle, *Politics* 1272a. The text is uncertain. The Law of Gortyn points rather towards a system like the Spartan.
113 The reasoning is the same as in the *Politics* of Aristotle: the sacred land is seen as taken from the communal land, since the cult is the affair of the community.
114 See e.g. Asheri 1966, 13–16. The oldest known description of this procedure is found in Homer, see p. 9.

situated in the same place, and finally have a glance at Sparta with her somewhat more dramatic history.

The only decree preserved concerning the distribution of private land in a new colony applies to Kerkyra Melaina (SIG^3 141). Kerkyra Melaina is an island off the west coast of ancient Illyria. At some time in the fourth century the citizens on the nearby, slightly smaller, island of Issa sent a team of colonists to the larger island, and at the same time the popular assembly of Issa carried a resolution which is preserved in part. Many modern concepts of the principles governing Greek distribution of land hail from this decree.[115] According to the decree, colonists were to be given land in two areas, one inside the walls and the other outside. In each of the areas they were to have a part of the choice or reserved land (*exairetos*)[116] as distinct from 'the parts' (*ta mere*) from which they shall also have their share.

Colonists arriving later were to be given a non-specified building lot inside the walls and in addition a well-defined unit of area from the outside which had not already been assigned to anyone (*adiairetos*).[117] The inscription concludes by mentioning the names of the first colonists and must have contained more than 200 names.[118] The decree ratifies an agreement between the citizens of Issa and two men, a father and a son, who are not Greek but are perhaps Illyrian potentates. As we are unaware of the specific conditions, we shall have to confine ourselves to using this as an example to show how the distribution of land to private persons *could* have taken place.[119] Obviously, the distribution of land is a matter for the state and, noticeably, a matter for the state from which the colony was founded. The principle is that the first colonists should receive lots of land of equal size. To ensure that the lots would in fact be of reasonably equal size, the land was divided into zones in advance, according to their location.[120] Probably, lots were drawn among the colonists concerning the plots of land of equal size within each zone separately. Not all land was distributed in this way, and from the surplus ground lots were measured out in equal size for the benefit of potential new arrivals.

As for the future regulation of the distribution of land, the decree only tells us that *one* small part[121] of the land allotted to each of the first colonists *must* remain in the possession of the family. One can guess at the considerations that, in this case, lie behind a provision like this. If the lots in question are located close to the city, it is possible that the purpose of this ban against

---

115 Especially Asheri 1966, 5–11, and, more nuanced, Asheri 1971.
116 *exairetos* usually means 'chosen by lot', Asheri 1966, 15, note 3.
117 This is the normal interpretation, but perhaps one should rather translate *adiairetos* as 'land which has not been set aside for other purposes' including land-lots for the colonists.
118 Brunšmid 1898, 2–14.
119 Cf. Finley 1968a, 28–30; for cities with part of the *chora* inside the walls see Martin 1973, 110.
120 The same principle has been used excessively in twentieth-century Greece, resulting in a fragmentation of the land, Thompson 1963, 23.
121 The text is illegible but it seems to be a question of not more than 1½ plethron.

selling such land would be to prevent a sensitive or indeed possibly vulnerable zone from ending up in just a few hands.[122] It might also be that, by this ban, the first colonists obtained a firm position in the corps of citizens for all time to come. This we cannot know without knowledge of the local criteria of citizenship. In his *Politics* (1319a), Aristotle maintains that in the old days many city-states not only had laws that fixed a maximum for how much land individual persons were allowed to possess, but also made provisions to forbid that anyone might ever sell the land that had first been allotted to him. Here the purpose is said to have been to increase the number of landowning citizens to a maximum, as it made for democracy in its most stable form.[123] In the Kerkyra Melaina decree it is envisaged that a new group of colonists would arrive. They, too, would have a right to lots of land, but the location of them would either be not quite so attractive or not quite so important strategically; there were no clauses with regard to transferability of land. Altogether, there is much evidence to indicate that, normally, private land was transferable in the Greek city-states.[124]

From Sybaris in southern Italy we hear about problems concerning the distribution of land between people who descended from the earliest colonists and new arrivals. Sybaris was founded by Greek colonists in the eighth century, but destroyed by people from Kroton in 510. Later attempts to restore the colony were in vain. In the year 444, on Athenian initiative, a new group of colonists was sent out; together with descendants of the Sybarites they were to found a new colony there. The colony was to be called Thourioi. The Sybarites felt that they were entitled to the best lots of land; their co-colonists, however, were unwilling to accept this claim. In the end, the Sybarites were expelled from Thourioi. Diodorus Siculus has it that the coveted lots were close to the city. Perhaps they were more fertile than others, or perhaps it was just a question of being close to a ready market. But it is also quite possible that the Sybarites did not like the idea of having lots close to the borderline because, time and again, their neighbours had demonstrated hostility. It would be dangerous to work there, and the risk of having their crops destroyed was considerable.[125] Once the Sybarites had been disposed of, the remaining colonists carried into effect an equitable distribution of land, or so Diodorus Siculus tells us. In Kerkyra Melaina land inside the city wall as well as land outside was to be divided up into portions to be distributed equally among the colonists. Archaeological investigations undertaken in various parts in the Greek area, particularly over the past twenty years, support the concept that this was common practice. As an example we may point to the investigation at Halieis where, in the sixth

---

122 For similar speculations concerning land at the frontiers see *IP* 3.12–14 where a foreign benefactor receives the right to own land in the territory of Priene provided that the land in question is situated more than 10 *stadia* from the frontier against the Ephesians.
123 Next to this came a democracy with a majority of cattle-breeders.
124 Finley 1968b.
125 Aristotle, *Politics* 1303a; Diodorus Siculus 12.11.1–2; Moggi 1987.

century, large, regular dwelling quarters were laid out. It would seem that the superior module was 50 plethra, precisely what investigations in quite different places indicate as having been used in connection with the parcelling-out of agricultural land.[126]

THE SELF-GROWN CITY

In the ideal states of the philosophers, as in the newly founded colony, there was full agreement between the number of citizens and the number of lots of land. For each set of lots, there was one citizen. This ideal situation could be maintained only provided the number of lots was allowed to remain constant; this would set a limit to the future number of citizens; so, that is what the philosophers did. Plato, in his ideal state number two (where, contrary to his first, he operated with privately owned land), decreed that only one son (one child) should be allowed to inherit the ancestral lot. Were there several sons, they would have to be distributed as adopted sons among households where there was no heir, and girls would have to be married off. Adjustments could be made by the expedience of sending out a colony or, if bad came to worse, by accepting a supplementary contingent of people from the outside (*Laws*, 739–40). Aristotle, on the other hand, was of the opinion that maintaining the distribution of land, agreed upon once and for all, would be preferable, this being accomplished by laying down legal limitations for how many children you would be allowed to give birth to (*Pol.* 1365a39–b14, 1335b20–7).

But none of the city-states in real life, which had been in existence for some centuries, could display an ideal agreement of this kind between the number of citizens and lots of land.

The first Athenian was born by Ge (the Earth) in Attica. That is where the Athenians had lived ever since – the city had not been founded by any single act. This is what the Athenians themselves have told us, and they are not likely to have told anything completely incompatible with what they observed round themselves.[127] Athens was demonstrably a self-grown city, and there is no reason to believe that even the most refined methods will ever reveal traces of an original division of Attic agricultural land into lots of equal size. Still, everything indicates that there was no major difference in the size of farms in Classical Athens.[128] It was a direct consequence of a consistent legislation with regard to citizenship and inheritance.

As we have mentioned (p. 118), particularly reliable sources are available in the speeches written by Isaeus concerning questions of inheritance in Athens, and among recent thorough discussions of the topic we may men-

126 Boyd/Jameson 1981, 328.
127 Cf. Loreaux 1981b.
128 See p. 79.

tion A.R.W. Harrison, *The Law of Athens* I (1968), and David M. Schaps, *Economic Rights of Women in Ancient Greece* (1979).

The number of Athenian citizens was fairly constant. By a law from 451 the number was controlled solely on the basis of criteria of birth: to become a citizen you had to have been born into a family consisting of parents who were both Athenian citizens. Sons of a couple of this kind could, once they came of age, be admitted to the list of citizens, while the daughters were the only women who could bear potential citizens and wives of such. Monopoly of the land belonged to the citizens; but there were far too many of them for everyone to possess land or to make a living from whatever they might have.

Owning land was not a condition for obtaining citizenship in democratic Athens; but by legislation attempts were made to prevent an amalgamation of existing properties, or *oikoi*. *Oikos* meant, literally, *a house*, but like our own word 'house' it had several connotations such as 'family', 'household' and in a wider sense 'everything a man owned'.[129] The laws applied to every kind of property, but they were no doubt phrased with a view to real property. Shortage of land and a political desire to preserve as many profitable farms belonging to citizens as possible would thus have contributed to the circumstance that, by law, Athenian women were defined as legally incapable of managing their own affairs, and consequently could not own land. It was customary for a dowry to be paid at the time of marriage, but this was usually paid in cash and therefore did not affect the distribution of land. Hence it follows, also, that the woman moved to her husband's house after marriage.

It was only if a woman had no brothers at the time of the death of her father that she might act as an intermediary of real property. She received the status of *epikleros* (heiress), and it was the duty of her closest male relative to marry her. Their first-born child, as soon as he came of age, was to take over the property of his maternal grandfather; if nothing else, the parents were under a moral obligation to have him adopted as a son of his maternal grandfather.[130] He thereby lost his right of inheritance with regard to his father's property. He was no longer a member of his own father's *oikos*. What was thus obtained, on the other hand, was that two houses of citizens could continue to exist without amalgamation.

The law governing adoption had the same function. Only a man who had no legitimate sons was allowed to adopt an heir. If the adopter had a daughter, it was a condition for adoption that the adopted son marry the daughter. In any event, at the time of adoption the adopted son[131] waived all

---

129 For a discussion of this cf. Xenophon, *Oeconomicus* 1, Aristotle, *Politics* 1, Finley 1973a, ch. 1, and MacDowell 1989.
130 This is the traditional interpretation of the law, cf. Fox 1985, 226. But see also Schaps 1979, 25–47.
131 Adoption of girls is testified in Isaeus 11.8–9 and 41. Both girls were daughters of sisters of the adoptee.

rights to the property of his own father; only provided he left a viable son in the *oikos* of his adopted father would he be allowed to apply for readmittance to his own father's house. Like all other laws, the law of adoption could be evaded, or it could be applied against its intention,[132] but this is irrelevant to our investigation. The law that brothers were to share the inheritance from their father on equal terms operated against the goal that each house should be self-supporting. In practical terms the high mortality-rate among children as well as the hazardous life led by young warriors had the effect that many persons would have to resort to adoption in order to have any heir at all.[133]

In Athenian legislation, out of consideration for the rightful heirs and with a view to maintaining the citizen's land, limits were laid down as to how much one was permitted to give away from a property. Purchase and sale of land was not prohibited and did take place to some degree.[134] It seems that, at least at the time of Theophrastus, there was a law in Athens with the provision that a contemplated sale of land (or of real property) was to be proclaimed before the authorities at least 60 days prior to the date when the transaction was to take effect.[135] Probably most people held on to their ground if they were at all able to do so, but land could also come up for sale in connection with confiscations.

The Athenians attributed most of the laws of inheritance mentioned above, like so many other laws, to Solon, and there is reason to believe that in this case we are in fact dealing with laws dating from his time.[136] In any case it may be observed that in Athens, since the time of Solon, no massive demand for a re-distribution of land (*anadasmos tes ges*) was ever heard; this was a demand that was otherwise, time and again, made elsewhere in the Greek area.[137]

In other words, the legislation proved expedient, although it resulted in a certain fragmentation of the land. The somewhat more affluent men whom we meet through inscriptions and speeches before the court were, typically, owners of land in different places.[138] Modern investigations on the peninsula of Methana have shown that fragmentation of this nature may be expedient, at any rate where the surroundings are dissimilar. It is only when you wish to mechanize and introduce artificial irrigation that the problems of fragmentation become acute.[139]

---

132 e.g. as a means to avoid liturgies or to insert a man illegally into the list of citizens, Isaeus 11.49–50; 12.2.
133 Isager 1981/2, 88–9. Burford Cooper 1977–8, 164–5. cf. also Fox 1985, 217.
134 Xenophon, *Oeconomicus* 20.22–6, Burford Cooper 1977–8.
135 Theophrastus, fr. 97 in Wimmer's 1866 edition; cf. Szegedy-Maszak 1981, fr. 21, with translation and commentary.
136 For a thorough discussion see Ruschenbusch 1966.
137 Davies 1977.
138 This is veiled if *chorion* ('a piece of land') is translated by 'farm' which is often the case.
139 Forbes 1976b, Thompson 1963; see p. 8.

SURPLUS LAND IN ATTICA

As we have seen, the Athenians had no re-distribution of land, neither in the time of Solon nor later. But they had an entire network of borderlines drawn over all of Attica in connection with the deme and *phylai* reform introduced by Kleisthenes in 507/6. Under the Kleisthenic reform, Attica was divided into 139 demes and not just 139 groups of men listed in the register of citizens. Most of the demes were defined as territories comprising an area of land with a main town or village.[140] The others, the city demes, were defined as sections in the city of Athens, but presumably without arable land of any particular consequence. Most demes already existed previously, but in certain places a mere hamlet was upgraded to become the main town of a deme. As was the case with the others, its territory would consist of private land, perhaps some sacred land, and the remainder (which, till then, must have been land owned by the state) now, unless otherwise defined, became deme land. The deme could procure income from this land either by selling it or by leasing it out.[141] In all probability, then, with the deme reform vanished almost all that may have been left of land belonging to the state, or common land in Attica.[142] New land might be added when property was confiscated, or when a border-area or an island was incorporated. Land confiscated within the limits of Attica, as a rule, was sold immediately. Border-areas like Oropos and islands like Salamis and Lemnos and perhaps Nea were never integrated into the Kleisthenic deme structure, so there it might be possible to find land owned by the state or common land to a greater extent. On the other hand, we see how, by a resolution passed by the popular assembly, the mountains of Oropos were divided among ten Attic *phylai*, thus making the administration of the land possible. In this case, it would make no sense to distribute the mountains among the smaller entities, the demes.

SPARTA

Like the Athenians, in the Classical period the Spartans had an official myth designed to legitimate their territorial claims. According to this myth, the Spartans had emigrated from the north and, by force, taken possession of the land that they now controlled; but they had a right to do so because they were descendants of Herakles who, once upon a time, had had Lakonia presented to him and thereby acquired the right to Messenia; by their arrival to the Peloponnese, as Heraklidai, they had merely collected their legitimate

---

140 Hans Lohmann believes he has discovered in south-western Attica a deme (Atene) with isolated farms but without a central town or village, Lohmann 1985; Lohmann in Wells, ed., 1992.
141 *SEG* 24.151 concerning Teithras, where the land of the deme is termed *ta koina* ('the common land'), here in reference to the community of the deme; cf. Whitehead 1986, 155, quoting Finley for the statement that the land of the deme was subject to private law.
142 Cf. Langdon 1985.

inheritance.[143] Myth may be useful when trying to convey the feeling of conditions in a certain area. The problem of Sparta is that it is difficult to reach a point beyond the myth.

Whereas it is possible, as we have seen, to form a reasonably precise picture of Athenian legislation concerning private land as a category, so many problems present themselves when we try to focus on the city-state of Sparta that some scholars doubt whether there was indeed anything there that could justifiably be called 'private'.[144] In the first place, this was due to the fact that Sparta, oligarchic and closed as it was, did not regard open government as a virtue in the Archaic and Classical periods. For instance, laws were not allowed to be laid down in writing, foreigners had difficulty in obtaining access, and the spiritual climate did not lend itself to literary enterprises. The result is that we lack contemporary written sources from within, and reliable written sources from outside Sparta. Epigraphic material with regard to the question of the distribution of land is practically non-existent. Among the ancient authors Xenophon is our main witness, providing useful information about internal conditions in Sparta during its days of glory. On the time following the renunciation of Messenia in the year 369, we have Aristotle's *Politics*, and from much later sources, apart from scattered remarks in the writings of Polybius, the biographies by Plutarch and the *Description of Greece* by Pausanias. It is a feature common to the authors mentioned that they themselves did not hail from Sparta, and only Xenophon had seen things from the inside during the period dealt with here. Plutarch as well as Pausanias wrote at a time when Sparta appeared in an archaizing form, almost like a tourist attraction. It is also unavoidable that Plutarch's sources are influenced by the Hellenistic reforms and attempts at such, under the kings Agis and Kleomenes, reforms for which a justification was looked for in the laws of Lykourgos.

The state of affairs with regard to sources has not prevented a constant and very comprehensive research on the history of Sparta. Today, we have recourse to Paul Cartledge, *Sparta and Lakonia* (1979), where archaeological findings and data from the modern Peloponnese are used extensively, but mainly concentrating on eastern Lakonia, and to Douglas M. MacDowell, *Spartan Law*, published in 1986, the same year that saw the appearance of S. Hodkinson's important article, 'Land tenure and inheritance in Classical Sparta'.[145]

The particular situation that applies to the sources presents us, for example, with problems of terminology. We cannot be sure that late authors of antiquity use terms that were technical terms in the Archaic and Classical periods. Nor, of course, can we take it for granted that they are aware of

---

143 Tigerstedt 1965, 28–36; Cartledge 1979, 76–7.
144 Among the sceptics are Pavel Oliva, cf. Oliva 1971, 32–8.
145 See also Hodkinson 1983.

conditions in Classical Sparta. Besides, the situation in Sparta was probably far from easily understood.

Whereas the Athenian territory measured approximately 2,600 km², the Spartans, during their period of glory – that is to say from *c.* 550 to 371 – were in control of no less than 8,500 km², the so-called Lakedaimon or *lakonike ge*, which Thucydides estimated to comprise two-fifths of the Peloponnese.[146] This large territory constituted a city-state with Sparta as its main city. But the *lakonike ge* was not the territory of the Spartans (Spartiates) in the same way as Attica was the territory of the Athenians. This has to do with the genesis of the city-state as reflected in the myth describing the homecoming of the Heraklidai; it also explains the differentiation in social status which provided for a small group of citizens to have more influence than others.

Within the area of Lakedaimonia there was a considerable number of city-states the citizens of which were known by a common name: the *perioikoi*; but they did, in fact, have their own individual names. You were a *perioikos* only as seen from a centre, and Sparta was so much so that the individual names of her perioecic states were of no interest to the outside world.[147] Each one of these states had their own area which they could presumably administer to their own satisfaction, that is, according to the laws issued by themselves. Yet the Spartiate kings had lots in the land of the *perioikoi*.[148] The land belonging to the perioecic states was not ideally suited for agriculture, but was nevertheless to a large extent arable.

Most perioecic states were located in old Spartiate territory, a few in Messenia which was conquered later in the sixth century.[149] Left for the use of the Spartans themselves was the Eurotas Valley and its surrounding country, and most of Messenia. This is probably what Polybius has in mind when he mentions *he politike chora* (6.54.3): the land available to the citizens.[150] We do not know what the Spartans called the different categories of land; but we shall have a look at the part that corresponds to the category 'private land', that which belonged to individual citizens.

At the time of Aristotle this land was distributed very unevenly. First, ownership was concentrated in few hands, and second, two-fifths of the owners were women, he says.[151] There is no reason to believe that Aristotle knew the correct ratio between land owned by men and that by women. But

---

146) Thucydides 1.10.2. This territory had been at least halved by the liberation of Messenia (369 B.C.) when Aristotle wrote his *Politics*.
147) Aristotle never writes about them. He regards the Helots as equivalents to the *perioikoi* of Crete. To him the neighbours of Sparta are the Argives, the Messenians and the Arcadians, *Politics* 1269b.
148) *exaireta*, Xenophon, *Lac. Pol.* 15.3.
149) For the location of the perioicic communities see Cartledge 1979, 185–93.
150) Cartledge 1979, 166–7; Hodkinson 1986, 385. For the opinion that we have to do with a technical term covering only the Eurotas Valley, i.e. the fertile land near Sparta, see Asheri 1961, 47.
151) Aristotle, *Politics* 1270a, 23–5.

there is also no reason to believe that he invented the main trend. This must mean that the Spartans had no legislation to secure the existence of a large number of profitable estates. What they did have was a law of civic rights which demanded that a full citizen should have his livelihood solely from agriculture and that, on the other hand, his land should yield sufficient for the particular individual to contribute his share to his *syssition*.

As the name indicates, the *syssitia* were common messes for citizens, and they were known in the cities of Crete as well. Here the adult citizen had his meals until he reached the age of 60, and in Sparta, at least, he was also expected to spend the nights there until his thirtieth year.[152] The common messes were a sensible result of the circumstance that the city-states were in a constant state of war; so the Cretan explains in Plato's *Laws* (625d).

At the time of Aristotle many Spartiates had been unable to deliver the stipulated dues to the *syssitia* and had, in consequence, forfeited their citizenship. This led to a state of affairs where there were less than a thousand full citizens left. It looks as if Aristotle would place the responsibility with Lykourgos, the legislator, who had neglected to legislate within certain subjects. The explanation of the absence of laws is sufficient and convincing. Unlike the Athenians, the Spartans were not so occupied by adjusting their laws.[153]

There is, however, a tradition which says that directly following their arrival in the Peloponnese, or at the latest in connection with the reforms introduced by Lykourgos, the Spartans divided their territory into 6,000 lots, a number soon raised to 9,000, of the same size, these to be divided among the ideal number of citizens (to wit, 9,000). As a phenomenon characteristic of Sparta, Plutarch and Polybius have asserted that all citizens possessed lots of equal size and that the number of citizens as well as the number of lots was kept constant.[154] Some scholars, then, have been of the opinion that all land was state property and that the citizens might be compared with government lease-holders whose lot would be handed back to the state at the death of the citizen, that is, the lease-holder;[155] others have thought that, admittedly, the land was state property, but that at the time of his father's death the eldest son took over the lot whereas a younger son, second in succession, would have to wait and see whether there was, or was likely to be, a vacant lot left in the pool of lots owned by the state.[156]

A concept of land in Sparta as being public land exclusively would agree

---

152 Plutarch, *Lycurgus* 15.7. The institution is discussed by Hodkinson 1983, 251–4.
153 Aristotle refers to a law stipulating that a man with three sons shall be exempt from military service, and from taxes too if he begets four. This is probably a late emergency law (Aristotle, *Politics* 1270b).
154 Plutarch, *Lycurgus* 8.3–4, 16.1; Polybius 6.45.3.
155 Among others, hesitatingly, Forrest 1980, 135–6.
156 Oliva 1971, 32–8, following among others Asheri. See Hodkinson 1986, 378. Figueira 1984 argues in favour of early allocation of lots to Spartiate male children and no inheritance of the lot of the biological father.

with the severe regulation of the lives of men by the state which, as we know, did take place,[157] but is not compatible with Aristotle's picture of conditions in the fourth century.

Plutarch holds a disagreeable fourth-century legislator named Epitadeos responsible for a law that resulted in an amalgamation of property and a decrease in the number of citizens; the law authorized the free transfer of real property by gift or bequest. This tallies well with the description of the administration of justice in Sparta that Aristotle gives us, as he knew it from his own time. For example, the 'legislator' (according to Aristotle) allowed for an *epikleros* to marry anyone. By choosing the term 'legislator', Aristotle is probably referring to Lykourgos, and at any rate this is not a new decree from his own lifetime that he describes. Aristotle does not mention Epitadeos.

We cannot know whether Epitadeos is a historic person.[158] Had there been two versions to choose between, Plutarch would have chosen the one where a person performs as an instigator rather than a version according to which the pitiable result was due to slow decay of the times. It is almost certain that, whatever the truth may be, an Epitadeos, if he ever existed, must have belonged to a period as early as the fifth century.[159]

It is possible that the discussion of the status of land has been based on the wrong premises. From the beginning, conquered land would have had the status of public land (*demosios*); but once the conquerors had taken their share, perhaps for the gods and surely for themselves, the citizens' land would have to be regarded as private (*idia*), even if there may have been restrictions concerning the transfer of land. This appears quite clearly from general deliberations in Aristotle's *Politics*; likewise it appears from his description of Sparta that this was his way of looking at the easy transferability of private land in Sparta, at any rate in *his* time.

Whether the Spartans had set certain land aside for public purposes when the distribution of land took place, we cannot know; but Aristotle's thoughts about the ideal state and his comparison of Sparta with Crete indicate nothing of the kind.

The myth describing the return of the Heraklidai was undoubtedly designed to explain why in Lakonia there lived thousands of the so-called helots, people with a status between that of slaves and free men; they were not counted as citizens but worked the land for the Spartiates (see p. 15).

It is part of the unclear picture of land conditions in Sparta that there must have been landowners who were neither *perioikoi* nor helots, but on the

---

157 Maybe already at birth, Plutarch, *Lycurgus* 16.1.
158 Asheri is undecided and sees 'il cosidetto "legge di Epitadeo" ' as part of a larger complex of fourth-century laws, some (unspecified) of which were only *de facto* laws, Asheri 1961, 68.
159 MacDowell 1986, 99–110, who regards Epitadeos as a historical figure. The reforms of Epitadeos are tentatively placed after 371 by Fox 1985, 221–2.

other hand had too little land to qualify as full citizens. As terms for persons of this middle class we find *mothakes*, *hypomeiones* and *neodamodeis*.[160] *Hypomeiones* ('inferiors') were perhaps mostly landowners who had arrived at a point beneath the requirements that allowed them to maintain a seat in the common mess.[161]

---

160 For a recent discussion of the terms see MacDowell 1986, 39–51.
161 Cartledge/Spawforth 1989, 42–3.

8

TAXES IN AGRICULTURE

> In a city-state . . . the land was in principle free from a regular taxation. A tithe or other forms of direct tax on the land, said the Greeks, was a mark of tyranny, and so firmly rooted was this view that they never allowed an emergency war tax, such as the Athenian *eisphora*, to drift into permanence (nor did the Romans of the Republic).

This is what Moses Finley writes at the beginning of his chapter, 'Landlords and peasants', in *The Ancient Economy* (1973a), 95–6. His postulate has turned out to be one of many useful dogmas on which one is forced to take a stand; indeed, others like Lewis, Roesch and Pleket have already declared that they are at variance on what was the traditional conception, prior to Finley.[162]

Whereas instances to illustrate Greek norms are customarily taken from Xenophon or from the court speeches and, reluctantly, from the contemporary philosophers, Finley, and before him Andreades, refers solely to Tertullian from the second century AD (*Apologeticus* 13.6). This means that contemporary, relevant, specific expressions concerning the attitude of the Greeks towards direct tax on agricultural produce and arable land are lacking, and that we have recourse solely to their practice. We shall have to define what is meant by the expression 'a tithe or other forms of direct tax on the land'.

Literally, a 'tithe' is a tax consisting of one-tenth of the annual crop, or at least some fraction thereof. '[O]ther forms of direct tax on the land', on the other hand, refers to a land tax based on an estimate of the value of the land. It goes without saying that the tithe does not require the same sophisticated type of society as does land tax, tithe being thought of simply as a certain part of the same kind of produce that is harvested, whereas land tax calls for an abstraction. It is not settled merely by delivering a cartload or two of earth, and so it often happens that a tax on land is found in states with a money economy.

---

[162] Like Finley but without special emphasis on the land Andreades 1965, 134–5. See Lewis 1959b; Pleket 1973; Roesch 1982, 287–98.

There is no doubt that in the Greek area in the Archaic and Classical periods tithe as well as land tax were well known. The question is how extensively the systems were applied, relatively and in absolute terms, and to which extent payments of this nature were viewed as a sign of tyranny. When trying to answer this question, we encounter difficulties with regard to terminology.

TERMINOLOGY

Greek vocabulary with regard to technical terms concerning taxation was not very elaborate, as was the case with regard to agricultural implements, and so on. The problem which, today, faces the scholar who attempts to understand the system of taxation is the circumstance that one and the same term may denote widely differing types of taxes – offerings to gods and taxes paid to the public purse, and voluntary gifts and compulsory dues paid by private persons. To this may be added the further difficulty in an agricultural context that few of the technical terms, by themselves, have agricultural connotations.

Plato employs the term *eisphora* to denote property tax (including land-value taxation) as well as tax on produce. The verb *eispherein* is also found, in specific cases, to denote the payment of poll tax.[163] In fact, all that is conveyed if we look at the etymology of the word is a reference to the act of paying in something, or making your contribution. As a rule the word is used solely about property tax.[164]

Duty on agricultural produce is usually referred to by the neutral word *telos*.[165] Otherwise words denoting fractions may be used, e.g., $1/10$ (*dekate*), $1/20$ (*eikoste*) and $1/50$ (*pentekoste*). None of these words may be said to have been associated with the sphere of agriculture only, nor to have been used exclusively in connection with dues to the state. This has to be determined, as a rule, on the basis of the context in each individual case. As for the term *aparche*, the situation is not quite so hopeless. One is able to determine that nearly always the addressee is a god, and for this reason *aparche* will be dealt with in the section on gods and the land. But, again, one must judge by the context whether a fraction of the agricultural produce is involved, or whether something entirely different is at stake, and also whether it is a case of voluntariness, statutory obligation or compulsion.

Fortunately, there were occasions when the Greeks availed themselves of circumlocutions. Concerning the Thasians, for instance, Herodotus uses the expression *ateleis karpon*, 'exempt from taxation on agricultural produce' (6.46). As a rule, however, only the philosophers take the trouble to explain what they understand by a certain term for tax or dues, owing to the fact, of

---

163 Pseudo-Aristotle, *Oeconomica* 1347a.
164 Sometimes the fraction of the property to be paid is indicated, Demosthenes 14.27.
165 For the Solonian *tele* cf. Skydsgaard 1988a.

course, that as against the reader or the audience they have to argue in favour of approval or rejection of the tax in question.

THE STATE IMAGINED

In his dialogue the *Laws* Plato has an Athenian, a Spartan and a Cretan debate the best way of organizing a new colony on Crete, provided, of course, that you had a free hand. Here Plato operates with two types of direct taxation, namely, property tax and tax on produce. The individual citizen would have to declare the value of his property whereas members of the *phylai*, on top of this, were to deliver a list of crops harvested, annually, to the magistrates called *agronomoi* (*Laws* 955d).[166] The authorities were then free to choose which type of tax should be brought to bear during that particular year. In the ideal colony of Plato's, direct tax should be levied every year (*Laws* 956d–e). According to Plato, property taxation is based on the total declaration of property for tax assessment and, therefore, not merely on the value of the land, although it must be assumed that in the ideal state, land constituted the major part of the citizen's property. Tax on produce was a variable fraction of the annual yield.[167]

Plato's imagined state by necessity differs from the contemporary city-states in many essential features; but he employs the concepts and terminology of his own time, and from his choice of words in general and his use of the term *eisphora* for property tax and tax on produce we may gather that both types of taxation were known as and could be called *eisphora*. We can know nothing about their extent, nor anything to show whether such taxes were regarded as a sign of tyranny. From Aristotle we gather that they were unpopular among the rich and likely to threaten the stability of the constitution if used in the most extreme democracies without revenues (*prosodoi*).[168]

TAXES ON AGRICULTURAL PRODUCE

A fixed duty on agricultural produce existed in Sparta where the full citizenship of the individual Spartiate depended on his ability to deliver his stipulated ration to his own mess, *syssition*. At the same time, at least as far as we know, this was the only collective duty that this tax on produce was to

---

166 The term used is *epikarpia*. Only what was left after contribution to the *syssitia* had to be listed (*choris ton eis ta syssitia teloumenon*).
167 The picture is blurred if the expression 'income tax' is used, as in Thomsen 1964, 44. That the harvest is in question is indicated *inter alia* from the fact that it appertained to the province of the *agronomoi*.
168 The *eisphora* is mentioned as the first among three dangerous means to provide payment for the political work in the democracy. The other two are confiscation and corruption of the law courts, Aristotle, *Politics* 1320a 20–2, both of which have been discussed by Aristotle in the preceding passage, 1320a 5–18.

cover, so that the Spartiates had to find other ways of covering further collective duties.

Dicaearchus, who lived in the fourth century and spent part of his life in the Peloponnese, part in Sparta, indicates the following (monthly) dues to be paid by the Spartiate as his contribution to his mess:

> approximately 1½ Attic *medimnos* of barley flour
> 11–12 *choes* of wine
> a certain quantity of cheese and figs
> approximately 10 Aeginetan obols to buy extras
> (Athenaeus, *Deipnosophists* 141)

Dicaearchus gives the ration in measures and in a currency known to his readers or his audience: an Attic *medimnos* corresponds to 52 litres, a *choe* to 3¼ litres. The accuracy of his information can only be guessed at, but there is no reason to doubt that he renders the principles guiding payment to the common messes. The first three items present no problems, but one may wonder at the amount of money since the Spartans were not in fact supposed to be in possession of money. Iron spits (*obeloi*) were the means of payment allowed, so Dicaearchus possibly merely assumed that the Spartiates themselves paid with iron spits.[169] Or perhaps what he meant was this: negotiable commodities corresponding to the value of 10 Aeginetan obols per month. When, in the first place, it was converted into monetary value, it could not be indicated in terms of the number of animals what a Spartiate was to pay per month. In practical terms it could well be that this part of one's dues was replaced by an animal now and then. Altogether, what the quote may convey concerning actual practice is limited.

Along with Cartledge (1979, 173), we may well wonder why olives are not mentioned. The fact that there is no direct agreement between what was consumed in the common messes and the dues paid would emphasize that a tax was involved.

Inasmuch as no definite fraction of the variable crop was fixed, but that this tax was represented by a monthly quantity per person from year to year, it was in fact a poll tax levied on the Spartiates. The distinction that existed between the Spartiates came to the surface, among other features, in that some could afford to spend wheat flour and other luxuries on their mess,[170] whereas others were forced to give up paying their dues and thus forfeited their status of citizen unless, as assumed by Fox, they bound themselves to a wealthier Spartiate who would then pay their dues on condition that he would inherit the land of the needy persons.[171]

---

169 Cartledge 1979, 173.
170 Xenophon, *Lac. Pol.* 5.3, Dicaearchus in Athenaeus, *Deipnosophists* 141.
171 Fox 1985, 222, with references. The conflict between the ideal of equality and the spurs to competition is discussed in Finley 1968a and in Hodkinson 1983.

On Crete, too, there was a permanent tax on agricultural produce, so Aristotle tells us, but here a better system existed. The main difference was that on Crete dues were delivered to a common pool (*Politics* 1271a and 1272a, cf. Plato, *Laws* 847e), whereupon a certain part was set aside for the gods and for state liturgies, whereas a second part was reserved for the *syssitia* (the common messes). In this way the individual ran no risk of losing his citizenship owing to his failure to pay to the *syssition*, and all men, women and children were provided for by the common means. Aristotle felt that this system was in better agreement with the idea of solidarity. He does not forget that there are several city-states on Crete (*Politics* 1269a–b), but he regards the system as being the same throughout the island; therefore, he speaks in general terms about the Cretans or about the Cretan constitution. His description is not entirely clear,[172] and very little evidence remains to support it. In any case, epigraphical material from a time as early as the fifth century, from Gortyn, one of the leading Cretan city-states, is available to substantiate the existence of a corps of *karpodaistai* ('distributors of harvest') whose duty it was to ensure that no part of the harvest had been put aside, and that everything had been divided. The fraction to be delivered is unknown. In the later Hellenistic period we hear about tithes (*dekate*) in connection with Lyttos as well as in connection with Gortyn.[173]

The Cretans and Spartans had a number of things in common. They were Dorian, or at least predominantly Dorian, oligarchies with common messes for the citizens as an essential element. This system was dependent on farmers who tilled the soil for the citizens. Had the dependent farmers of Sparta been asked, their answer would probably have been that their duty to deliver part of their produce to the Spartiate whose land they tilled was a sign of tyranny. However, the citizens imposed upon themselves the further deliverance of part of this to the community as a fixed and direct tax on produce, a sort of ticket of admission to their own privileged circle.

Dues on agricultural produce were not limited to states of the Spartan or Cretan description. Herodotus tells us about Thasos that here the citizens were *ateleis karpon* ('exempt from tax on crops'). It is evident that this struck him as something out of the ordinary, and as the reason for this exemption from taxation he gives that the Thasians had a great number of other state incomes, first and foremost from their mines.[174] This was probably a decisive factor in determining whether there were taxes on agricultural produce in a Greek city-state.[175] Another important factor was the question of sources of supplies.

Sparta as well as Crete was amply supplied with grain for its own

---

172 Some editors emend the text in 1272a 16–18 which describes the source of the contributions to the common pool.
173 *IC* iv 77; i 18.11; iv 184. Athenaeus 4.143B = *FGH* 399, Dosiades about the Lyttians.
174 Herodotus 6.46. *IG* XII, suppl. 349 from the end of the fifth century indicates that the Thasians by then paid regular taxes on their agricultural produce, Salviat 1986, 152–3; 181.
175 Pleket 1973.

purposes,[176] but for most city-states the problem was how to provide sufficient supplies of grain and other provisions for the population. Like the majority of city-states, Athens had no surplus of agricultural products, possibly with the exception of olives. There were few large estates, but many smaller ones. The difference between Athens and the majority of city-states was the fact that Attica had an exceptionally large number of inhabitants, and that many citizens were not landowners.

Offhand, therefore, there is reason to believe that Athens, with her various other sources of income, had no permanent state tax on agricultural produce, as well as reason to look for a different way to explain the cases of doubt that do in fact exist.[177]

In a newsletter from the American School of Classical Studies at Athens, T. Leslie Shear, Jr, reports that in the course of excavations of the Agora in 1986 a marble stele was uncovered; its inscription records a law proposed by Agyrrhios in Athens in the year 374/3, 'Law on taxation of the twelfth part[178] of the island grain'. It is said to be evident from the text itself that the islands concerned are Lemnos, Imbros and Skyros, and that the dues consist of payments in kind, wheat as well as barley. It is specified how much each tax collector is to bring back to Athens. The law also specifies how, and by whom, the grain should be collected, where and when it should be put on board ship and transported to Piraeus, stored in the agora, and then sold. The tax collectors receive, among other things, a certain percentage of both kinds of grain.

The inscription is too important to be allowed to pass unmentioned although some of the questions occasioned by the newsletter are undoubtedly mere pseudo-questions that will be solved automatically once the inscription has been published. Lemnos, Imbros and Skyros were Athenian cleruchies. According to current concepts, until now, this means that the land belonged to Athenian citizens.[179] It seems probable, as suggested in the newsletter, that the law owes its existence to a shortage of grain in Athens.[180] It is grain which is needed, not, for example, the equivalent amount in silver. If we are in fact dealing with a law and not a decree, the idea was, presumably, that the dues were to be permanent. At first sight it is surprising if the cleruchs were to pay a special tax that other Athenian citizens did not pay. On the other hand, we know that owners of cleruchic land were in

---

176 Although Gortyn gets part of a large gift of grain from Kyrene in 330, when there was a general lack of grain. Tod 1948, no. 196.
177 *Pentekoste*: Andocides 1.133; Demosthenes 59.27. *Dekate*: Meiggs/Lewis 1969, no. 58.7, cf. Lewis 1959b, 243–4.
178 $8^{1}/_{3}$ per cent according to Shear; $^{1}/_{12}$ in connection with *eisphora* in Demosthenes 14.27. In the ideal colony of Plato they divide the harvest contributed in twelve portions, $^{1}/_{12}$ for every month, *Laws* 847e.
179 Gauthier 1966, 67; 1973.
180 See also Garnsey 1988a, 147.

some respect privileged as compared with owners of land in Attica.[181] So the tax could be a counterweight to the privilege.[182]

But until Agyrrhios' law has been published, we must confine ourselves to acknowledging that with it we have yet another separate law applying to an area which at one time was part of the Athenian territory, but which was never integrated into the deme–phyle structure. Another such law is that pertaining to Nea, an area that was, until recently, tentatively identified with Oropos; now, following Merle Langdon's suggestion, it should perhaps rather be identified with the island of Nea which was situated between Lemnos and the Hellespont.[183]

TAX ON LAND

As we have mentioned, Plato uses the term *eisphora* to denote taxation on property as well as on produce; but when we are dealing with actual city-states *eisphora* practically always refers to property tax.[184] Property tax in real life, as was the case with Plato, was calculated on the basis of the total property of the individual; therefore, there was a variance as to which part of a certain *eisphora* was made up from land taxation.

There is no specific Greek term for land tax, but this does not mean that it was not customary to evaluate land. As usual, our best information stems from conditions in Athens. There, land lots were evaluated according to their value in terms of money[185] when they were to be used as security for a loan, when it was a matter of taking over the responsibility of administration of the capital on behalf of a fatherless minor or when the matter at issue was taking over the bride's dowry. The land was also evaluated in connection with *antidosis* when a dispute was at hand as to who would have to undertake a liturgy, and it was evaluated at sale and at leasing. Thus tax on land was paid together with tax on other property if and when *eisphora* was imposed. The individual landowner was required to estimate the value of his land and add the amount to whatever else he might have in the way of capital. Part of the land belonging to the Athenians was never taxed. In the

---

181 Clerouchic land could not, in the mid-fourth century at least, give rise to trierarchic obligations, Demosthenes 14.16, cf. Gabrielsen 1991, 125–9.
182 More tempting is the thought that no tax at all is involved, but a compulsory delivery of grain for which those who delivered received a certain remuneration, cf. Plato, *Laws* 848a. It could be considered a collateral to the law that prohibited the export of grain grown in Attica. Compare also the law of 350 on transport of ruddle from Keos, Tod 1948, no. 162. Keos was not Athenian but only a member of the Second Athenian Confederacy.
183 Pliny, *Naturalis Historia* 2.89. Stephanus of Byzantium, s.v. *Neai*, cf. Langdon 1987, 56 note 28.
184 For the expression *eispherein* used about an extraordinary tax, which also implied a poll tax on persons without property see pseudo-Aristotle, *Oeconomica* 1347a 18–24 concerning Potidaia.
185 Not so in the Archaic period, since the highest Solonian 'property class' was called *pentakosiomedimnoi*, referring to the size of the harvest.

first place, many small landowners were not in possession of so much property that they were required to pay tax. Cases of tax evasion, however, were not unknown. The splitting-up of land, which we have mentioned previously, combined with the absence of an official cadastre, would at times result in the fact that only the owner himself had a clear picture of how much land he in fact owned here and there. The Athenians distinguished visible from invisible property. Visible property was what the owner was known by others and/or acknowledged himself to possess. It would as a rule consist of movables, but there are indications that land could also be included in that category.[186]

As pointed out by Finley, *eisphora* was, when possible, considered an extraordinary measure. In Athens the only accepted reason for imposing it was, for a long time, that it was essential for military purposes; this was an attitude that Xenophon, in his *Poroi*, in 355, agitated against. It seems that in Athens, at the latest from 349 (up until 323), it was imposed each year in order for the arsenal at Piraeus to be built. In Mende the *eisphora* was, in a way, permanent. It was known how much the individual taxpayer was to pay annually in *eisphora*, or rather what he should have paid. It was preferable to allow the taxpayer to keep his money as a state loan free of interest, and manage the daily public finance administration with money derived from customs duties and the like. 'Taxes from land and houses' were not retrieved until it was absolutely necessary, but under those circumstances were retrieved with retroactive effect. After all, the persons involved were in arrears (pseudo-Aristotle, *Oeconomica* 1350a 7–12).

The Spartan city-state was always short of means, Aristotle tells us; his explanation is that the politically powerful also owned the majority of the land and were, therefore, at the same time potentially the most important taxpayers. Admittedly, they could well afford it, but few as they were, they were also in a position to reach an agreement by which control of each other's payment of taxes was omitted.[187]

In contrast to Athens and Mende, Sparta was a society, in principle, without money.[188] The Spartans willingly accepted contributions to war from abroad, also in coin;[189] but, when the citizens of Sparta were prepared to come to the aid of others from abroad, they could not have recourse to their treasury. Instead, they would fast for a day, and would not feed their animals, it was said. Probably Plutarch felt the same urge as we do to understand how the fasting of the Spartans could be converted into assistance, and his version of the story gives us details: the rations of food and

---

186 Gabrielsen 1986.
187 Aristotle, *Politics* 1271b, cf. Thucydides 1.80.4.
188 See p. 138.
189 Meiggs/Lewis 1969, no. 67 about support to Sparta, probably under the Peloponnesian War. It is not certain if the *perioikoi* had to pay tribute to the Spartiates.

fodder for humans and animals respectively were collected and, in this particular case, sent to the starving citizens of Smyrna.[190]

Direct tax on land, then, among the Greeks existed as part of property tax. The example from Mende seems to indicate that this was an ordinary tax. When protests against it were heard, the protest did not imply that land was something special, something upon which no tax should be levied. It was aimed at property tax in its entirety, no matter which type of property was involved. Direct taxation on the produce of the land was brought to bear where resources and ideologies permitted.[191] The Spartiates did not in any way regard it as degrading that they were to surrender part of their crops to the community. On the contrary, it was degrading if you were not able to pay.

It is one matter when the citizens of the city-state impose a tax on land or crops upon themselves for the benefit of the community. It is a different matter when a tyrant levies a similar tax. From him, no consideration could be expected with regard to the situation of resources nor to the wishes of the community. The Athenians had experienced it with Peisistratos – or at least, so it was told in the city (Aristotle, *Ath. Pol.* 16.4). In the Hellenistic period, tax, especially on crops, was an entirely accepted method for the kings to secure quick and safe income from the remotest corners of their realm. The Romans adopted the system, and the distance between taxpayer and the official who imposed the tax increased steadily. Tertullian had lived to see this development. In his time the circumstance that direct tax on land or crops was to be levied was undoubtedly *signum captivitatis*, as he puts it.

INDIRECT TAXES ON AGRICULTURAL PRODUCTS

Purchase tax was a common type of indirect taxation in the Greek city-states, but naturally it served no purpose to have it where no business was anticipated. For this reason, there is no talk about purchase tax on agricultural products in Plato's ideal colony. This was to be self-sufficient with regard to agricultural products (although, from the point of view of the state, it was necessary for artisans who were not landowners, as well as foreigners, to purchase the part of the agricultural products set aside for them). The other two portions would be distributed direct to citizens who would then look after their *oiketai*.

In all this, it is no coincidence that his ideal colony reminds us of Sparta, where indirect taxes on agricultural products should not have been an issue, inasmuch as native crops were not objects of trade, and the state was

---

190 Meiggs/Lewis 1969, no. 67; pseudo-Aristotle, *Oeconomica* 1347b about help to Samos; Plutarch, *Quomodo adulator ab amico internoscatur* 64B about Smyrna, cf. Isager 1988, 81.
191 For a discussion of examples that seemingly contradict Andreades and Finley see Andreades 1965, 161–70.

reasonably self-sufficient regarding agricultural products.[192] The Cretan system as known to us also resulted in a very limited trade in agricultural products so that the possibility of a purchase tax did not apply to them.

The other extreme point is Athens where there was a general purchase tax including tax on agricultural products; business was necessarily considerable. But comparatively few of the agricultural commodities sold on the market were actually grown in Athens; Athens needed a large import, first and foremost of grain.[193]

Customs duties also belong in the picture in so far as transactions across borders were concerned. This immediately excludes Plato's ideal colony as well as, in most respects, Sparta and the Cretan city-states; but in Athens there was a general duty of 2 per cent on all imports and therefore, also, on imported agricultural products.[194] Duty on grain was one of the major items of income in the state budget and had to be paid by the importer, no matter whether he resided in Athens or in the Crimea. Differential treatment is known, however, from places where a dynast was in power. For example, the kings of the Bosporus exempted ships transporting grain to Athens from export duty, in return for which, by virtue of an honorary decree, they were allowed *inter alia* to hire *hyperesia* ('petty officers for their war ships') in Athens.[195]

---

192 But see Chapter 10.
193 This remains true even if the calculations of Garnsey in favour of a high productivity in Attica are accepted, cf. Garnsey 1988a, 89–106, but see p. 26.
194 A 5 per cent tax on all import replaced the tribute in 413, Thucydides 7.28.4.
195 Tod 1948, no. 167. *Hyperesia* in lines 59–60 is understood as 'petty officers' by Morrison 1968, 254–6; Gabrielsen 1991, 156.

9

OTHER LAWS

COMPULSORY CULTIVATION

In Athens you could receive a summons to appear in court charged with *argia*, often translated by 'laziness'. In ancient days this crime was punished by death, but in the Classical period there was a fine of 100 drachmas for the first two offences; in the case of a third ruling by the court the penalty of *atimia* was incurred. It is uncertain what constituted the crime, as the word *argia* is ambiguous, and no speech made before a court dealing with this type of trial has been preserved. According to the way the law was interpreted by an orator of the fourth century, the issue was the fact that idleness was forbidden (Demosthenes 57.32), but since this does not tally with what we otherwise know about the Athenian attitude towards work, it has been suggested that to omit trying to bring oneself out of poverty was prohibited.[196] But it seems that we are in fact dealing with an old law stipulating that it was everyone's duty to cultivate his land. This is the interpretation favoured by Theophrastus who thought that it was introduced by Peisistratos with a dual purpose, partly for the purpose of having land cultivated to a greater extent or more intensively, and partly so as to get people away from the city (Plutarch, *Solon* 31.5). Others ascribed the law to Dracon or Solon (Plutarch, *Solon* 31.5. Harpocration, s.v. *eranizontes*). Either would be understandable. The law may be viewed as a parallel to that which stipulated that it was not allowed to squander one's paternal inheritance (Aischines 1.94–105). Or it could be a codicil to Peisistratos' law which stipulated a duty of a fixed fraction of the agricultural produce (Aristotle, *Ath. Pol.* 16.4).

REGULATION OF SUPPLIES

In Athens a duty was also levied on export of commodities, but among the agricultural commodities of Athens, only olives were affected. By law, all

---

196 'habitual idleness', Hansen 1973, 80–8, and 1976, 73.

other crops were not allowed to be exported.[197] A prohibition of this sort was scarcely unusual. We hear of such prohibitions in connection with the city-state of Selymbria, on the coast of Thrace. On one occasion when there was a great shortage of grain in other Greek areas and a shortage of money in the treasury of Selymbria, but a surplus of grain, the surplus was exported, but under the control of the state: the farmers were to hand over to the state everything beyond what was needed for a year's consumption at a price fixed by the state. Then the state offered the same grain for sale to whomever wished to export it. Now the price was set according to the demand in the districts struck by famine. In other words, the state imposed an extraordinary purchase tax.[198] There is, however, at least one example of an embargo on an agricultural product: Thasian ships were not allowed to import foreign wine to Thasian territory.[199] One might imagine that this was an anti-luxury law like those Plato wanted for his Magnesia; but Thasos produced her own exquisite wine in great quantities, wine which was known throughout the Mediterranean area. When an import ban on Thasos was maintained, it may have been in order to prevent mixing an inferior foreign wine with Thasian wine whereby the latter ran the risk of losing its high reputation.[200] This is a protection of business interests, although, like minting, it cannot be entirely dissociated from concepts like honour and dignity. Several such examples are known but until more instances have come to light this law must be regarded as an exception to the rule that, primarily, consumers' interests and fiscal considerations were the dominant motives for the regulations governing import and export in the Greek city-states.

In the fifth century, owing to their position as leaders of the Delian League, the Athenians found it in their power to direct supplies for Athens. This was not the case in the fourth century, and around 350 BC the so-called 'maritime suits' were introduced; they were directed towards quick decisions in lawsuits when maritime trade on Athens was involved.[201] In particular, it had to do with a noticeable improvement of the legal rights of captains and merchants, so it is clear that this legislation was caused by considerations of supplies as well as for fiscal reasons.

OTHER STATE RULINGS

In the preceding pages we have mentioned certain important instances where agriculture was regulated by law. An entire small code of laws for farmers

---

197 Plutarch, *Solon*, 24. Isager/Hansen 1975, 35.
198 Pseudo-Aristotle, *Oeconomica* 1348b–49a. We follow the reading of Van Groningen, contrary to Garnsey 1988a, 75.
199 IG XII, suppl. 347 II; Salviat 1986. The written sources on Thasian wine are now collected and discussed in Salviat 1986.
200 If foreign ships could bring non-Thasian wine to Thasos, the effect of the ban is questionable, cf. Salviat 1986, 183–7.
201 See Isager/Hansen 1975, 55–87.

may be found in Plato's *Laws* 842e6–846c8. What is at stake here are disputes that may occur between neighbours, and Plato makes no claim to any great degree of original thinking. Rather, the section reflects his wish that the ideal colony should function in practice:

> This has been adequately stated by many lawgivers, whose laws we should make use of, instead of requiring the Chief Organizer of the State to legislate about all the numerous small details which are within the competence of any chance lawgiver.
>
> (Plato, *Laws* 843e–844a)[202]

A detailed perusal of Plato's laws on farmers where each individual ruling is compared with those which we know, by coincidence, from various Greek city-states, has been undertaken by Eberhard Klingenberg in his *Platons nomoi georgikoi und das positive griechische Recht* (1976). Reliable bits of information from the city-states are miles apart from each other, except when it does not specifically concern sacred land; but whenever information is available, it serves to support Plato's own words. To a large extent he has used laws that were already operative; but the way legal proceedings were applied could, of course, be at variance with that of the real city-states where there were also differences in procedure.

As an example we may refer to the question of utilization of natural water-courses, a question that is of essential interest in areas with a shortage of water, and where rain, when it does fall, is often a torrential downpour which is likely to cause more damage than the good it might have yielded.

In Plato's *Laws* a case is anticipated when a farmer wished to lead water on to his land. This is permitted provided he takes the water from a public water-course (*Laws* 844a). A similar ruling is known from an inscription from Gortyn: a law specifying that no more water may be taken than an amount that ensures that there will remain a sufficient supply of water in the main tributary, the public river (*IC* IV 43B).

Rain-water can create a problem for those who own land on mountain slopes. With Plato, in general, consideration is advocated, and disputes in such cases are referred to the magistrates called the *agronomoi*. In a speech before the court in Athens, from the fourth century, it is seen that a claim for damages could in fact be the result of a dispute concerning the diversion of rain-water. The scene of the drama is the slope of a mountain where the participants in the dispute were opposite neighbours, their fields being divided by a public road.[203] The charge was that, by constructing a ditch or a fence round his land, the defendant had blocked the former conduit where rain-water used to flow like a mountain-stream, thereby leading the rain-

---

202 Translated by R.G. Bury, Loeb edn.
203 Demosthenes 55.23–4. For another interpretation of the situation see Osborne 1985a, 17–18.

water on to the road from where it flooded the plaintiff's field and caused damage to his crop. In other words, it was unlawful to block a water-course of this kind in such a way that it would cause damage to others. In a very eloquent apology the defendant denies that there was ever a proper brook through his land;[204] on the contrary, without any protest from his opposite neighbour, this land was fenced in already by his father. Former owners had buried their dead there, and the land had long been planted with fig and vine. It is quite normal and legal to discharge water from one's land on to the road, but – he claims – this is in fact not what he has done. Had he allowed rain-water to flush down uncontrolled upon the fields of his neighbour lower down, the latter could rightfully have complained. Then, what was he to do with the water? 'Would you have wanted me to drink it?' he asks the plaintiff Kallikles; the judges must have had an enjoyable day!

Once again, we have an inscription from Gortyn that provides us with a law concerning water, probably rain-water (*IC* IV 73A).[205] Here it is made clear that it is not unlawful to lead rain-water on to a neighbour's field, provided the latter, in the presence of witnesses, has declared that he agrees.

We have not mentioned any laws concerning animal husbandry. In Plato's Magnesia laws are laid down that concern anyone who lures swarms of bees to his own premises or who allows his cattle to graze where they have no business (843d 6–7). Nothing similar is known as far as bees are concerned, and the only laws about cattle prohibit them from sacred areas in existing city-states. This does not necessarily mean that such legislation did not exist.

From Knossos of the fourth century fragments of an inscription are preserved; there, damages are fixed which one has to pay if one has broken the horns of an ox belonging to someone else; in the same fragment we read that one who has purchased a beast (*kartaipos*) is eligible to claim the deal void if, within five days, he has returned the animal and paid the 3 obols per day, which was the price for the use of a beast of burden of that description (*IJG* 19a).

---

204 It has to be *homologoumene* to be considered a proper creek, Demosthenes 55.19.
205 Guarducci (1935–50) *ad locum*; Klingenberg 1976, 106.

10

LABOUR AND STATE

Xenophon, in his *Oeconomicus*, lets Socrates sum up what he and Kritobolos have agreed on as being the best occupation for 'a handsome and good man', and Socrates concludes as follows:

> We came to the conclusion that for a gentleman[206] the best work and the best science is agriculture, from which men obtain what is necessary to them. For this work seemed to be the easiest to learn and the pleasantest to work at, to give to the body the greatest measure of strength and beauty, and to leave to the mind the greatest amount of spare time for attending to the interests of one's friends and city. Moreover, since it makes the necessary things grow and nourishes them outside the walls, agriculture seemed to us to stimulate in some measure those working with it to become able defenders. And so this way of making a living appeared to be held in the highest estimation by our states, because it seems to turn out the best citizens and most loyal to the community.
>
> (*Oeconomicus* 6.8–10)[207]

In other words, the city-states should prefer that its members, the citizens, are farmers out of consideration for upkeep, their physical and psychological constitution and finally their will to defend the territory. In fact, most of the laws of city-states known to us presuppose that the land is owned by the citizens. On the other hand, the laws never demand an identity between landowner and farmer. Even as in Sparta, a landowner could be prohibited from tilling his own land. In this section we shall have a closer look at the type of farmers known from Greek territory, and at the legal framework that governed their labour.

---

206 i.e., 'the handsome and good man'.
207 We do not quite follow the translation by E.C. Marchant, Loeb edn.

STATE AND AGRICULTURE

THE BEST STATE

In democracies everybody deals with everything. The more a city-state tends towards oligarchy, the greater will the splitting-up of necessary functions become, Aristotle concludes. In his own ideal state, not particularly democratic, by necessity the citizens are landowners, but not farmers. In their youth they are occupied by serving as soldiers; later on in life they take over as persons who are legally competent to make decisions; and then, perhaps, end their days as priests. However, they need considerable amounts of supplies, and for this they depend on their own land. Tilling is to be done preferably by slaves or possibly by those whom he calls barbarians, who are *perioikoi*. His state is a figment of fantasy; but a system whereby farmers and soldiers operated separately still existed in his own time on Crete and in Egypt, so Aristotle claims (*Politics* 1328b–1329a).

CRETE AND SPARTA

As we have seen, it is from Aristotle and also from Plato that we hear about the presence of a separate group of people on Crete who tilled the land for its owners. When, at the end of the last century, in Gortyn, an inscription was found with a code of laws stemming from the fifth century (*IC* IV 72), this population group could be identified as the group which, in the inscription, was called *oikeis*, or on rare occasions *douloi*. The inscription shows that each *oikeus* had a master (*pastas*); it also shows that, for instance, he had a right to the children who might be the result of an alliance between his *oikeis* (cols iii, 52–5; iv, 18–23). On the other hand, an *oikeus* could have his own cattle, and the house in which he lived could not without further notice be vacated or taken from him upon the death of his master (col. iv, 31–6).

The same passage in the inscription shows that *oikeis* lived in the country. This probably had to do with the fact that they had no business in town because they were not citizens. On the contrary, their presence away in the fields was a prerequisite for the citizens, the landowners, to remain within the city where they would be able to participate daily in the *syssitia* (the common messes), as well as exercising in athletic sports and the use of arms. Aristotle does say that, wisely, the Cretans treated their farmers so well that the only difference between them and the free men was merely the circumstance that the dependent farmers were not allowed access to the gymnasia and could not possess arms;[208] typically, the gymnasia were located inside the city.[209]

In the Gortyn legislation on inheritance, houses in the city are mentioned separately. Sons were to inherit the house, or houses, in advance, along with other property, before the inheritance was distributed among sons and

208 Aristotle, *Politics* 1264a 19–25.
209 Traders and craftsmen also lived in the town but in special quarters.

daughters, in the relation 2:1 (col. iv, 29–43). In exceptional cases an *oikeus* could marry the daughter of a citizen. Were he, in such an alliance, to move in with her, their children would be free men, that is, they would become citizens. Were it the other way, the children would have the status of their father. The former would probably happen only if the girl turned out to be an heiress and found no man within the circle of persons who *must* be asked. Perhaps one might go one step further and venture a guess that the *oikeus* in question would move into town and thereby join the circle of citizens even if his citizenship would not become valid until the next generation (col. vii, 1–10).[210] This assumption would agree with David Asheri's slightly provoking statement that by definition, in Greek territory, city-dwellers were looked upon as colonists, whereas people who dwelt in the countryside were regarded as the original population, regardless of historical facts.[211]

It is likely that dependent Cretan farmers, at least according to their own beliefs, were those who lived there before the arrival of the Dorians on the island. A similar case applied to those among the farmers who tilled the soil for the Spartiates. Those who cultivated the land around Sparta had had the status of helots ever since the arrival of the Dorians, whereas the Messenian helots had not been subdued until the final conquest of Messenia in the seventh century. Since in Sparta it was not customary to commit laws to writing, let alone have them hewn on stone, there is no hope of finding anything corresponding to the Gortyn law there. For this reason, it is likely that we shall forever be ill informed about the details concerning the legal status of the helots. The helots, too, lived in the country and could have their own families. It is likely that they could be neither sold off by the master for whom they worked, nor removed at random; but either could happen in the case of a state decision. They belonged to the community and each of them had a special responsibility towards a specific member of it. Their number is not known, but it was much higher than the number of Spartiates.

Whereas, apparently, the dependent farmers on Crete paid the due on their produce direct to the state, the individual helot was liable to duty towards the Spartiate whose land he cultivated. He was to surrender a certain part of the produce, half of it as claimed by Tyrtaeus (fr. 6), whereas Plutarch may be interpreted to the effect that it represented a fixed quantity per year and that it consisted of 82 *medimnoi* of barley and the equivalent of fruits (*Lycurgus* 8.7; 24.2; *Moralia* 239d–e).[212] In the same place, he notes that it was unlawful to demand more than the fixed amount. Both statements present a problem, that of Tyrtaeus because he wrote at such an early time and besides may not have meant to have his statement taken literally, and that of Plutarch because he wrote at a much later time.

210 The term used in the text is *doulos*, not *oikeus*.
211 At the congress 'Agrigento e la Sicilia Greca: Storia e Immagine (580–406)', May 1988.
212 So, hesitatingly, MacDowell 1986, 32–4.

We cannot assume that the individual helot and the family that he might have, in the Classical period at least, had land of the same quantity or quality to cultivate; therefore it is not likely that everyone was to deliver the same quantity to his master. So, if in fact there was a law common to all helots, it is most likely that they were to deliver a certain fraction of the crop, the size of which is unknown to us.[213] Many questions remain unanswered because we do not know how many helots there were in relation to the number of Spartiates. For instance, what happened if a lot of land could no longer provide for the helots as well as their master? On the other hand, how did the Spartiate deal with the large surplus of grain and other commodities of which he had become the owner once he had acquired a great deal of land?

As distinct from the Cretan *oikeis*, the helots constituted a permanent threat to their masters and to the state as such, as is seen from their attempts at rebellion, and actual rebellions that were occasionally successful. Aristotle explains the difference between the two communities by assuming that, in general, the Spartans treated their dependent farmers worse than the Cretans did, and by stating that all the Cretan city-states had the same type of dependent farmers and for that reason would not dare support an attempt at rebellion among those in a neighbouring *polis*. Sparta, on the other hand, was surrounded by states that by themselves had no helots, and for that reason could find their advantage, given the opportunity, in supporting the Spartan helots in a rebellion (*Politics* 1269a–b). The everlasting state of war between landowners and farmers in Sparta was emphasized by the fact that each year, on behalf of the state, the ephors formally declared war against the helots; furthermore, this was part of the education of young Spartiates, literally to decimate the number of helots by way of nightly assassinations (Aristotle, fr. 538).

Thus on Crete as well as in Sparta there was an entire group of underprivileged whose function it was to cultivate the land for the owners, and according to Aristotle they find their parallels in the *penestai* of Thessaly (*Politics* 1269a).[214] As a warning against taking Aristotle's description of contemporary societies too literally, it may serve to remember that he makes no mention of the *perioikoi*, well attested through other sources in connection with Sparta. Aristotle reserves the term *perioikos* for the dependent farmer on Crete, those who, in the Gortyn law, are called *oikeis* and correspond to the helots of Sparta. This terminology corresponds to that which was applied by his older contemporaries, Isocrates and Plato, to whom *perioikoi* are groups of people who live in a state of dependence, a state that in some cases may be described as tantamount to slavery (Isocrates, *Panathenaicus* 178; Plato, *Laws* 547c3). This might indicate the reason why Aristotle

---

213 For the advantages of share-cropping and for a wide and thorough discussion of agricultural labour see Jameson in Wells, ed., 1992.
214 Contrariwise, the so-called Aristotle fr. 586 in Photius s.v. *kallikyrioi* should hardly be considered a genuine quotation from Aristotle.

does not mention the *perioikoi* as a special group – that they lived in a greater degree of dependence than is generally assumed today. Xenophon's famous description of the prelude to the uprising in Sparta in 395 points in that direction. An informer is about to reveal the plot to the Spartiates and says that all helots, *neodamodeis*, *hypomeiones* and *perioikoi* are let into the secret: 'for whenever among these classes any mention was made of Spartiatae, no one was able to conceal the fact that he would be glad to eat them raw' (*Hellenica* 3.3.6).[215] Another reason why Aristotle does not mention the *perioikoi* in Lakedaimonia may also be that he considers them peripheral and not integrated into Sparta, for which reason it would merely distort the general picture to bring them in.

ATHENS

In Athens the law stated clearly that only citizens could own land, but in contrast to the city-states we have dealt with so far, not all citizens were in fact landowners. There was no particular order of farmers and consequently no specific legislation concerning a group of this status.

The democratic constitution did not demand that by necessity the citizen must find himself inside Athens at all times in order to fulfil his duties as a citizen, and there was definitely no full identity between city dwellers and citizens. Living together, mostly in the city and in villages (see p. 69), there were representatives of all three orders of society: citizens, metics and slaves.

The farmer was known as a *georgos*. This term tells us nothing about his status but indicates only that he worked tilling the soil. Xenophon's Ischomachos is a citizen and a landowner; but Xenophon awards him the title of *georgos* primarily because he understands the art of agriculture and takes an active part in farming, yet first and foremost as if he were a commanding officer. Those who worked for Ischomachos correspond to the soldiers; occasionally, Xenophon has chosen to describe them as *ergatai* (5.16.1–2), a neutral term that may also be applied, for example, to builders. This expression tells us nothing about the order of the persons involved but it does tell us something about their status by virtue of the fact that they work for someone else. In some places, Xenophon calls them *douloi* or *oiketai*, whereby it is indicated that these farmers were in fact slaves. The foreman was called the *epistates*, another neutral term, and only the context shows us that he, too, is a slave.

There would have been no reason why the slaves of Ischomachos might not also have been known as *georgoi* (cf. 3.10.1–3; cf. Aristotle, *Politics* 1329a). In inscriptions from about 330 BC, discovered on the Acropolis, containing lists of freedmen and freedwomen who have presented a silver bowl to Athene upon being given their freedom, the word *georgos* to denote the

---

215 Translated by C.L. Brownson, Loeb edn.

profession of the freedman occurs frequently.[216] After having been given his freedom, this person would receive the status of metic; consequently, he cannot have been a landowner, but could have been a farmer. Athens had known dependent farmers before the reforms of Solon, but with his *seisachtheia* and ban against debt-bondage, that group disappeared. After that, either the farmers were owners tilling the soil with their families only, or the owner could supply his labour force with chattle-slaves or with day-labourers who would normally be men but occasionally might also be women. Finally, the farmers could be leaseholders who in their turn could supplement their own labour with slaves or free day-labourers.

The question of the extent to which slaves were used as labour in Athenian farming is not clear and cannot be definitively answered owing to the nature of evidence. Jameson argues in favour of a very considerable extent in his article, 'Agriculture and slavery in Classical Athens' (1977/8); Ellen Meikins Wood disputes this point of view in her article (1983) and in *Peasant-Citizen and Slave* (1988).[217] Wood, however, attaches a greater and a different role to leasing than previously assumed.

LEASING

Within the limits of the law, there was room for leasing in several city-states. This, for instance, applies to Athens where, unfortunately, information in this respect is very sporadic. Our best information concerns the leasing of land belonging to the gods; our sources concerning the leasing of any other land are extremely limited. As far as land owned by the state, or public land, is concerned, this in all likelihood indicates that no public land of any consequence was available for leasing. We cannot, however, attempt to make the same deduction where privately owned land is concerned. Inscriptions containing private agreements were usually set up only when one of the parties was not present and thus unable to protect his own interests. This is the picture reflected in the *horoi* preserved in Athens (Finley 1973c, first edn 1952); it is also from them that, indirectly, we acquire some knowledge of the leasing-out of privately owned land in a certain type of case; this occurred when the lessee was fatherless and not of age. The guardian of the fatherless child could choose to have the inheritance of his ward leased out in its entirety until the ward came of age. The leasing-out was to take place via the popular court, and appraisers appointed by the state were to ensure that whoever received the lease was able to present a guarantee in real property. In order to protect the property of the heir who was not yet of age, *horoi* were placed on the land which was provided as security; from this we learn that lessees of the property of the fatherless children were citizens, and that they were already

---

216 *IG* II/III² 1553–9 with new fragments added by Lewis 1959a and 1968. Commented on by Jameson 1977/8, 133–5.
217 For further discussion see Jameson in Wells, ed., 1992.

owners of real property. *Horoi* do not indicate to what extent, in each individual case, land was part of what had been leased, and only occasionally remarks on individual cases turn up in a speech before the court. Robin Osborne (1988) argues convincingly in favour of the assumption that a considerable amount of privately owned land had been constantly leased out in this way in Attica.

The main source for the leasing-out of land by private persons is Lysias 7 (the speech on the sacred olive-stump), where we hear about the vicissitudes of a piece of land. It had been confiscated from an enemy of democracy and presented to a metic, Apollodoros from Megara, who had earned the gratitude of the state by participating in the murder of another of those revolutionaries (Lysias 13.71). Two years later he sold it, and the new owner leased out the plot, but subsequently sold it to the man now charged with having removed a sacred olive-stump from the piece of land in question. He himself had leased out the land for a period to a former slave (Lysias 7.4–5; 9–10). These brief passages from Lysias are of great relevance because they tend to show that leasing of private land was a common phenomenon in Attica. Unlike the lessees of the property of fatherless children, the former slave was not himself a landowner. Most likely the farmers, *georgoi*, mentioned in the lists of freedmen from the Acropolis, leased the land they cultivated just like him. Whereas it was the *entire* property of the fatherless child, an economic entity, that was leased out to those who were already landowners, it is possible that the piece of land mentioned by Lysias was not large enough to support a family, and therefore was leased as a supplement to other income.

Ellen Meikins Wood may well be right in assuming that it was precisely among lessees of privately owned land belonging to fatherless children that the poorest farmers were found, those who served as labourers for landowners on land that was not so conveniently located as the rest of their property.[218]

Leases are, to our knowledge, not indisputably attested in Classical Crete and Sparta, and presumably leasing played a minimal role, if any, there. Inasmuch as leasing is predominantly well attested and frequent in connection with sacred land, it will be dealt with more thoroughly in our section on gods and agriculture.

---

218 Wood 1988, in agreement with Osborne 1988.

Part III
GODS AND AGRICULTURE

INTRODUCTION

The gods were masters of the weather, and thus responsible for the growth of plants. Often, the gods themselves were landowners; they could also own cattle. They received a large share of their offerings in the form of agricultural products. For that reason, this section will discuss how the individual farmer and the city-state as a community and agricultural society administered their relation to the gods.

With this approach we shall have to venture into the domain of the history of religion where, however, we shall refrain from discussing intricate problems such as the relation between myth and ritual, questions concerning the age of festivals and many other relevant topics.

The situation concerning our sources has already been described in part in the Introduction to Part II (see p. 117). It should be emphasized that the epigraphic material is extraordinarily rich where the relationship between agriculture and gods is concerned.

We cannot entertain a hope of treating the subject exhaustively, but wish in this section to suggest some variations and, in particular, some constant features, which are numerous because all Greeks had their pantheon more or less in common.

11

THE CALENDAR

There are many good discussions of the Greek calendar system. Among general works of reference we may mention E.J. Bickerman, *Chronology of the Ancient World* (1980), whereas the Attic calendar of festivals has been especially thoroughly studied by L. Deubner, *Attische Feste* (1932, 1966) and in a lighter vein by H.W. Parke, *Festivals of the Athenians* (1977). At the time when Parke was writing, J.D. Mikalson was preparing a study entitled *The Sacred and Civil Calendar of the Athenian Year* (1975); it is a survey of the Attic year, day by day, indicating what we know about the individual day with regard to state festivals and days when councils or popular assemblies were to take place. It turned out that days of festivals that recurred annually were kept free, as far as possible, for meetings of councils and assemblies, whereas the monthly days of festivals were only kept free for meetings of popular assemblies. For obvious reasons, the relation between days of festivities and work in general could not be shown by the investigation. But for the ideal we may listen to Aristotle, who when dealing with associations within the city-state writes that many of these

> combine to perform sacrifices and hold festivals in connection with them, thereby both paying honour to the gods and providing pleasant holidays for themselves. For it may be noticed that the sacrifices and festivals of ancient origin take place after harvest, being in fact harvest festivals; this is because that was the season of the year at which people had most leasure.
>
> (*Nicomachean Ethics*, 8.9.5)[219]

The connection with agriculture is emphasized by Erika Simon in *Festivals of Attika. An Archaeological Commentary* (1983), and this connection is the main subject of Allaire Brisbane Chandor, 'The Attic Festivals of Demeter and their Relation to the Agricultural Year' (1976). To her, as to us, Martin P. Nilsson's *Primitive Time-Reckoning* (1920), like his many other pioneering works, is still of great value.

219 Translated by H. Rackham, Loeb edn. For harvest festivals see p. 166.

THE CALENDAR

AGRICULTURAL YEAR AND AGRICULTURAL CALENDAR

There is no great difficulty in producing a chart to illustrate the agricultural year in Greek territory in antiquity (Figure 11.1). The basis for this chart is primarily the agricultural calendar found in Hesiod's *Works and Days*, 381–617. Admittedly, it lays down limits for the possibilities of drawing general conclusions because the poem was presumably composed in one location, Askra in Boeotia, and the calendar makes no claim to be complete (see p. 7). Hesiod's didactic poem may be supplemented with the help of Theophrastus and by more coincidental sources concerning the year of the farmer. Last, but not least, we may seek support in investigations of agriculture as practised in more recent periods.

One cannot, however, draw a calendar for individual estates. Not all of them had all the crops; our chart includes only the four vital ones. On most estates, cattle-breeding was probably of little or no importance. All we have shown in the chart is the time when, according to Sophocles, the sheep went to the mountains. Most of the fields probably lay fallow every other year. Grain was frequently grown on the field where olive was also planted. The latter means that when ploughing your olive grove, you ploughed your field of grain at the same time. These were not two separate operations, and the growth of the grain determined when you could plough your olive grove. The chart merely indicates the period within which a job was done, but it does not show how long the job lasted. Therefore, there is a good reason why the chart should have been placed only after the section on agricultural technique.[220]

It is, and always has been, of vital importance for the farmer to pay attention to time. Otherwise hunger or dependency threaten. The question is how, in antiquity, you went about determining when the time was ripe. In our modern times, with satellites and television, it can be difficult to envisage how the reckoning of time was performed in Greece before the beginning of our era. However, it is perhaps easier to imagine how they managed where the agricultural calendar is concerned. The farmer's working year never followed any official calendar, but rather 'the natural year', the year in agreement with the solar cycle and hence the change of the seasons. The farmer looks at signs in nature to show when the time has come for ploughing and sowing. To the Greek farmer, it could be the arrival of a certain migratory bird. In Aristophanes' comedy *The Birds*, in agreement with the topic of the play, the birds boast of their enormous importance to the farmer (*Birds*, 709–15). In the chart we have merely indicated the part played by astronomical signs as heralds of seasons for the beginning of an agricultural piece of work. The farmer was to note the appearance of the

[220] For the agricultural year in antiquity see also the commentary of West 1978 on Hesiod's *Works* ll. 381–517 and the appendix; compared with the evidence of building inscriptions in Osborne 1987, 15 (see p. 104). For the modern agricultural year see McDonald/Rapp 1972, 51.

Grain	Vine	Olive	Varia	Signs			Demeter (Kore)	Dionysos	Zeus (Hera)	Athene	Apollo	Artemis	Other gods
			*Harvest of figs	Sirius rising	Hekatombaion	July				Synoikia Panathenaia	Hekatombaia		Kronia
	Vintage				Metageitnion	August					Metageitnion		Herakleia
Ploughing Sowing		*Harvest and pruning begins	Woodcutting	Arcturus rising	Boedromion	September	Eleusinian Mysteries			Niketeria	Boedromia	Artemis Agrotera	Genesia Demokratia
				Pleiades setting	Pyanopsion	October	Proērosia Thesmophoria	Oschophoria	Proērosia Apatouria	Apatouria Chalkeia	Pyanopsia		Theseia
				Orion setting	Maimakterion	November			Pompaia				
(Late ploughing)				Solstice	Posideon	December	Haloa	Rural Dionysia					Posidea
					Gamelion	January		Lenaia	Gamelia				
Ploughing of fallow field	Pruning finished Opening of jars			Arcturus rising at dusk	Anthesterion	February	Chloaia Lesser Mysteries	Anthesteria	Diasia				
			*Sheep to the mountains		Elaphebolion	March		City Dionysia				Elaphebolia	Asklepieia
Harvest	Digging finished			Pleiades rising	Mounychion	April		Olympieia	Olympieia		Delphinia	Delphinia Mounychia	Bendideia
Threshing Ploughing of fallow field					Thargelion	May				Kallyntheria Plyntheria	Thargelia	Thargelia	
				Solstice Orion rising	Skirophorion	June			Dipolieia Diisoteria	Arrephoria			
						July	Skira						

*Not mentioned by Hesiod.

Figure 11.1 The agricultural year and the Attic calendar of festivals

constellations of the fixed stars on the night sky (their heliacal rising) and their disappearance. Solstice was another important fixed point in the agricultural year, but naturally it could not be observed directly like the constellations of the stars. Summer solstice coincided with the rising of Orion which lends itself to direct observation. Nothing similar was to be seen at winter solstice, but the Greeks had their natural or manufactured sighting-marks. On Crete a stele from the fourth century has been found bearing an inscription which runs:

> Patron set this up for Zeus Epopsios. Winter solstice. Should anyone wish to know: off 'The little pig' and the stele the sun turns.
>
> *(IC IV.11)*[221]

The little pig was a rock formation way out in the water. The stele was not found *in situ*, but must presumably have been placed so as to show that when the sun was seen rising or setting in the sighting-line between those two points, then it was winter solstice.

THE SACRED CALENDAR

To Hesiod and to his contemporaries the celestial bodies were gods or at any rate divine beings; but they were not gods to whom you prayed or offered sacrifices. 'Dem Menschen, seinen Sorgen und seiner Frömmigkeit bleiben solche Machten fern', Burkert writes.[222] Sun and moon were constant and there was no special reason why one should try to remain on good terms with them. All you had to do was to abide by the signs they and the stars gave, just as one had to observe the arrival of the migratory birds and certain indications in the wild fauna. With their help time was measured.

This does not mean that the farmer could manage without the help of the gods; when autumn ploughing was to begin and the grain was to be sown, it was wise to offer sacrifices to Zeus and to Demeter as well. They were the two gods who were of prime importance, Zeus because he sent the rain, and Demeter because she was responsible for the growth of the grain more than any other god. Experience showed that even if the farmer had sown at the right time, Zeus could delay the rain or send a rain so heavy that it would be necessary to undertake a new sowing which could be a serious and perhaps insurmountable obstacle. In other words, it was of vital importance to sow at the right time and also offer sacrifices to Zeus.[223]

Hesiod wrote for the benefit of the individual landowner or farmer who

---

221 *SIG*³ 1264. For summer solstice see Plato, *Laws* 945e. For the measuring of time with sundials, *polos* and *gnomon*, see Diels 1965, 155–60.
222 Burkert 1977, 271–3, agreeing with Nilsson.
223 Hesiod, *Works* 465. Xenophon, *Oeconomicus* 5.18–20; 17.2. As is his habit, Xenophon does not specify the name of the gods.

GODS AND AGRICULTURE

would then, as later on, naturally present his private offerings in accordance with the agricultural year. So his advice did not become outdated. But as soon as the society became more complicated, and the inhabitants of a larger area acquired common interests, the necessity of a more specified calendar made itself felt – among other reasons so as to ensure that the gods might receive the common offering from the community at the right time. That, at least, is how it is explained in Plato's *Laws* (809d):

> What I allude to is this – the arranging of days into monthly periods, and of months into a year, in each instance, so that the seasons, with their respective sacrifices and feasts, may each be assigned its due position by being held as nature dictates, and that thus they may create fresh liveliness and alertness in the State, and may pay their due honours to the gods, and may render the citizens more intelligent about these matters.[224]

The month, too, played its part by the time of Hesiod, but there only as a single entity.[225] There is a section at the end of his work which deals with the nature of the different days of the month (765–821). It is stated:

> For these are days which come from Zeus the all-wise, when men discern aright.
> To begin with, the first, the fourth, and the seventh – on which Leto bare Apollo with the blade of gold – each is a holy day. The eighth and the ninth, two days at least of the waxing month, are specially good for the work of man. Also the eleventh and twelfth are both excellent, alike for shearing sheep and for reaping the kindly fruits.[226]

Here, no month name is mentioned.[227] The nature of the days is the same each month. Therefore, we cannot see if there was in Boeotia an official calendar of the moon which he was able to refer to. Some specialists in calendary systems and in research on Hesiod have been of the opinion that it cannot be Hesiod who composed the section on the days of the month because it reflects a slightly later stage of development and a mentality different from that in other parts of the poem.[228] To us it makes no great difference whether it was in fact Hesiod because if not, we are still dealing with an almost contemporary interpolation, and we aim at an overall picture, not just a picture of Hesiod.

As long as it is merely a matter of doing the right things at the full moon,

---

224 Translated by R.G. Bury, Loeb edn.
225 Only one month name is mentioned, namely Lenaion, l. 504. For the view that the passage is an interpolation see Bickerman 1980, 99, note 26. Lenaion is not a Boeotian month. Claire Preaux sees a conglomeration of several chronological layers in *Works*, Preaux 1973, 79.
226 Translated by H.G. Evelyn-White, Loeb edn.
227 There are no month names in Homer either.
228 Cf. West 1978 *ad loc*.

the month is well suited to dividing the larger entities of a year – for instance, the season during which you can sow. A proper calendar becomes available only once an attempt is made to connect the lunar year with the solar year, as was the case in various places in Greece in the young city-states of the seventh century. Possibly under Babylonian influence,[229] they all had a year consisting of twelve lunar months, and most of them, for instance Athens and Delphi, had the year begin at midsummer, ideally from the first new moon after solstice. In Boeotia, on the other hand, the new year was counted from the first new moon after the winter solstice, whereas Miletos and her colonies seem to have used spring equinox as the dividing-line.[230] The fact that new year as such was not really celebrated in the Greek city-states is borne out by this confused state of affairs. The important thing was for the individual state to have a point of adjustment.

The names which the Greeks gave to their months differed from city-state to city-state; but they all named them after one of the festivals for the gods of the particular month.

THE FESTIVAL YEAR IN ATTICA

It is only when dealing with Athens, democratic and keen on documentation and festivities, that sufficient material is available to reconstruct the festival calendar of the state (see Figure 11.1). As far as the period before the introduction of democracy is concerned, we can only guess, because of the sources at hand, although clearly many of the festivals are older not only than democracy, but also than the city-state as a whole. The notion of a state festival is artificial and not unambiguous, and perhaps it is of greater interest to observe that it is, in fact, ambiguous. One might say that the state festivals were those for which the city-state of Athens was responsible, contrary to, for example, the deme festivals.[231] In that case the rural Dionysia should not have been included in our calendar, for they were celebrated in the manner of the deme and only in some of the demes. The festivals took place in the month of Posideon, but not at the same time in the individual demes. There is nothing to testify that there were very many meetings in Athens during the month when the rural Dionysia were celebrated. This is one sign (among others) indicating that the festivals were of great importance to 'all Athenians', and for that reason we have included them in our chart.

The first eight days of each month, except the fifth day, were already dedicated to certain gods before the introduction of the calendar, and in Athens they were officially celebrated each month. We have disregarded these festivals as we are occupied with the rhythm of the whole year.

Festivals for Apollo have given their names to most months in the Attic

229 Nilsson 1951.
230 Roesch 1982, 33. Ehrhardt 1983, 120–2.
231 Cf. Whitehead 1986, 176–8.

calendar in spite of the fact that these festivals were of lesser importance in Athens, at least during the Classical period. This seems to us to suggest a foreign element in the monthly calendar.[232] No doubt, it was better suited for some place other than Athens.

The festivals for Demeter, the grain goddess *par excellence*, were placed at critical points in the farming year. Her festival year may be interpreted simply as the year of the grain. This does not apply to any other gods, but several of the other gods did have certain festivals that were also related to the cultivation of grain and fixed at times that corresponded to the Demeter festivals. There is no official harvest festival for Demeter,[233] but the Thargelia celebrated for Apollo and Artemis may be interpreted as a state harvest festival. To judge by the rituals, so could the Kronia celebrated in honour of Kronos two months later, when masters joined slaves for revels. This is also the way it was explained in antiquity, but the festival occurs very late in relation to the harvest and was therefore better suited on Samos and in Perinthos where the month of Kronion, during which the Kronia occurred, coincided with the Skirophorion of the Athenians. In Athens, the Kronia would appear to have been celebrated as a new-year festival.

Dionysos was god of the wine and in his honour the Oschophoria were celebrated when the wine had been harvested, and the Anthesteria when the wine was ready for consumption. The vine required patient care all year round, but the stock lasted for many years, and for that reason, among others, it did not lend itself to the rhythm of the festival year in the same way as grain. Another reason is that its importance for the sustenance of the Greeks is less than that of grain.

Olive, the third main crop of the Athenians, was protected by Athene. So far, it has been assumed that growing of the olives had no special festival of its own. Erika Simon, however, presents a convincing, or at any rate tempting, argument to show that the Arrephoria for Athene and Aphrodite in midsummer was a festival to secure the dew that was necessary during the months from then on until the time of harvest if the fruits were to grow to an adequate size.[234]

As Athene's special tree, the olive played an important part at the Panathenaia where the prize was oil from Athene's sacred olive trees. This festival was one of the greatest of the year and had no specific relationship with agriculture; it was a sort of national festival. It took place in the first month of the year. That was the time when patriotic sentiments ran high along with the festival celebrating the synoikism of Athens.

There followed a month nearly free from festivals after which came the

---

232 For a recent contribution to the discussion see Simon 1983, 73–6.
233 Harvest festivals were probably held privately and in honour of Demeter, Chandor 1976, 147–52.
234 Simon 1983, 9–17, underpinning and developing further the theory set forth by Deubner 1932, 9–17.

month of Boedromion which may be said to have been divided into two halves. In the first part, emphasis was of a military nature. The end of the warfaring season was celebrated with the Niketeria for Athene, with a memorial festival for the Marathon heroes, celebrated in the name of Artemis, and a memorial festival for those who fell during the season and for the dead of the family, the Genesia. Democracy was also celebrated. In practical terms war could be waged throughout the year, but it was propitious to strike while the enemy's grain was either ripe on the fields or newly harvested so that you could either forage or burn off the enemy's reserves for the next year.[235] When waging war in Metageitnion, on top of that, you could also collect the enemy's crop of wine or have it destroyed. As far as the navy was concerned, this was the time to have the men-of-war hauled on land. Consequently the end of the warfaring season was placed officially in the calendar with festivals in the first half of Boedromion.

In the second half of the month, the Eleusinian Mysteries prepared for the new agricultural year. There are no Attic festivals which by themselves may be characterized as agricultural festivals. The Thesmophoria, for instance, may also have to do with human fertility, and there are many things in the ritual that elude our understanding. Even if we were to find their original significance, it is likely that in the Classical period the Athenians changed their emphasis, and that furthermore each participant had her own understanding. The military year contributed to establishing the festival calendar of the Athenians, but as we have seen, it was dependent on the agricultural year. The festival calendar of the Athenians confirms Parker when he states: 'In a farming community the emotional year, as it may be called, is shaped around the agricultural year.'[236]

THE CALENDAR FRIEZE IN ATHENS

Next to the cathedral of Athens there is a small Byzantine church, Hagios Eleutherios, the façade of which is decorated with a frieze at the height of the building where, on Doric temples, a sequence of metopes and triglyphs was placed. The frieze on the church is ancient and taken from an unknown, 'pagan' context that we cannot reconstruct. In its original position the frieze was divided into two parts. At the time it was moved it turned out that the parts were slightly longer than the façade of the church. Therefore, part of each of the two parts of the frieze were chopped off, and the sequence of the parts was inverted, with the result that today, if you wish to arrive at the original message, you must begin reading from the middle. (In addition, the Christians have embellished the decoration by cutting Maltese crosses on to the frieze at regular intervals.)

---

235 Hanson 1983, 30–5. The conflict of interests is stressed: the soldiers abroad would be missed at home if there were no Helots or the like to take care of the harvest.
236 Parker 1983, 29, with a reference to Durkheim.

The frieze is our oldest well-preserved pictorial representation of a Greek calendar.[237] No one can tell how old it is; but it is probably from the Hellenistic period.[238] We are dealing with a calendar frieze. The individual months are personified as young men, and one is left in no doubt that they represent the Attic months. The rhythm in the presentation is set:

1 the personified month;
2 allegorical representation of one or more important festivals during the month, or the depiction of an important aspect of agricultural labour; and
3 the constellations of the months.

The connection with agriculture is evident. In its original state, the frieze began with Pyanopsion, corresponding to our October–November. This was the beginning of the season for ploughing and sowing, the time of year with which Hesiod started his description of the cycle of the agricultural year. Festivals which are directly associated with agriculture are prominent on the frieze, although it also shows the important Panathenaia; and in Maimakterion, an area poor in festivals, we see, instead, a man ploughing and behind him a sower.

The arrival of a new season would be marked by a personification at intervals of three months and in the same place as that shown in the agricultural calendar of Varro,[239] that is, autumn in the sign of the Lion, winter in the sign of Scorpio, and summer in the sign of the Bull. Spring must have been on the part of the frieze now lacking.

The calendar frieze displays an intended attempt at unifying the natural year with the official calendar, based on lunar months. It does not fit, but the viewer cannot see it. The frieze cannot have served as a practical calendar, neither for farmers nor for those whose duty it was to make arrangements for the festivals of the month. It does, however, provide us with the rhythm of the festival year of the Athenians, dominated as it was by festivals having a more or less direct association with agriculture. Newer festivals such as the two annual celebrations for Asklepios and for Bendis have not been included, and one also searches in vain for the Apollonian festivals that lent their names to the months. We cannot know which monument the frieze adorned in antiquity; it was probably of a sacred nature. The lack of a reliable dating limits its value as a source to us, but the fact that the frieze was used on the Byzantine church must be an indication of how the church absorbed some of the most persistent concepts from antiquity.

237 A red-figure badly fragmented krater dating from about 375 BC depicts five months personified as young men, each with a new moon over their head and holding an attribute referring to an important festival of the specific month. The festivals are the same as those depicted on the calendar frieze, Simon 1965.
238 Deubner 1932, 248–54, with tables 34–40. Simon 1983, 6.
239 Varro, *De Re Rustica* 1.28, cf. Skydsgaard 1968, 43–63.

12

AGRICULTURAL PRODUCTS FOR THE GODS

There are three reasons for offering to the gods – the wish to honour them, gratitude and the need for a benefit; so Theophrastus writes in his work on piety (Porphyrius, *De Abstinentia* 2.24). At first sight, these three reasons might seem to correspond to three types of offerings. But in a description of Greek sacrificial practice no such clear-cut categorization according to motive is feasible. The Greek practice of offering was not built on a logical, theological system, as noticed by Ziehen in his article 'Opfer'.[240] Instead, although not enthusiastically, Ziehen chooses a mechanical categorization of the sacrifices according to the type of object offered. Burkert, in his investigation of the origin of the rituals, emphasizes the basic difference between offering from the first-fruits on one hand and offering from slaughter on the other.[241] Even if this may have been the case in its origin, the offering of the first crop is no longer so easy to reduce to a formula in the period dealt with here. Nevertheless, for practical purposes, we shall adhere to Burkert's distinction because it may serve to illustrate important aspects of the offerings of agricultural products.

FIRST CROP OFFERING

When in his *Symposium* Xenophon has Socrates ask his friend Hermogenes what constitutes his 'divine service' since he is on such good terms with the gods, Hermogenes answers that he returns to the gods some of all that they have given him (*Symp.* 4.49). Xenophon does not allow Hermogenes to use a technical term for his sacrificial ritual, but what he describes is a first-crop offering.[242] The Greek term for this type of offering is *aparche*, which means 'from the first'. Thus the word itself has no agricultural connotation, unlike the English 'first-fruits'.[243] Characteristics of the *aparche* were that it was

---

240 Ziehen 1939, 581–2.
241 Burkert 1979, 52. Isocrates, *Archidamus* 96, imagines how the Spartiates would be ashamed to see their (former) *oiketai* give richer *aparchai* and *thysiai* than they themselves from the land they once received from their ancestors.
242 Mikalson 1983, 21.
243 Contrariwise, the less-used *thalysia*, from *thallos* ('a young shoot').

something normally given only to gods, or sometimes to the dead; that it was part of, or at least represented part of, something greater; and that this part was offered so as to show one's gratitude at having received the whole. Whenever the expression is used more generally, these must be the associations aroused.

Burkert mentions that the decisive characteristic of the *aparche* was its being offered as something that you gave away to the god, consequently something that would not be eaten if articles of food were involved, but that might be thrown into the fire or left to rot or dry up. As we shall see, this characteristic is not always at hand in the period with which we are dealing. As *aparche* you might offer inedible things such as the hairs from a sacrificial animal, part of your spoils of war, or perhaps a statue which had been bought for this share (Demosthenes 12.21). But frequently it was precisely articles of food that were involved, a casual part of a crop or a certain fraction of it. Sometimes this offering was named from the portion that supposedly constituted part of the whole. Therefore, it often happens that *dekate* ('a tenth') is used synonymously with *aparche*.

As in most other relationships concerning gods and agriculture, the phenomenon of the *aparche* is encountered in the private and semi-private domain as well as on an official level. The offering made by Hermogenes, mentioned above, was an entirely private undertaking. In connection with the Pyanopsia for Apollo, groups of children would go from door to door in Athens with a branch of olive or laurel decorated with newly harvested fruits (*eiresione*), which they placed before the door while singing an ancient song, expecting the occupants to give them a little something in return. A similar branch was placed before the temple of Apollo, and in the course of the same festival a mixture of all sorts of crops, *pyanopsion*, was offered; this lent its name to the festival as well as to the entire month. In these cases the ritual or symbolical significance was probably considerable, but the quantitative significance measured in terms of privation on the part of the donor or in terms of gain on the part of the god in question was negligible. Thus, the olive branch was allowed to remain and dry up until the following year when it would be replaced by a new one.

It is a different matter when in their capacity as administrators of the shrine of Demeter in Eleusis, the Athenians compelled by law all Attic landowners and all landowners in the Athenian empire to deliver an *aparche* to Demeter of Eleusis (*IG* 1³, 78).[244] To all Greeks who belonged to neither of these categories, an exhortation was to be issued with a view to a voluntary contribution of first-crop offerings in Eleusis. The law cannot be dated with any degree of precision but was enacted in the latter part of the

---

244 In fifth-century Athens there was no clear distinction between laws and decrees. We would call the enactment in this case a 'law' since it was meant to be permanent, cf. Hansen 1978.

AGRICULTURAL PRODUCTS FOR THE GODS

fifth century.[245] Every Athenian or ally was to deliver 1/600 of the harvest of barley and at least 1/1200 of the crop of wheat.

The inscription shows that we are not dealing with an entirely new phenomenon. In former times it had been customary to deliver first-crop offerings in Eleusis, and the Athenians had also received an answer from the Delphic Oracle to support them (ll. 4–5). In fact, such offerings to Demeter are perhaps attested by an inscription datable as early as approximately 460.[246] We cannot tell the rules according to which offerings were made before the law referred to above, nor whether a system did in fact operate. But we can see that hereafter the arrangement was to become permanent: the only thing that might vary from year to year would be the time of delivery. Naturally, this would depend on the time of harvest.

Initially, however, the offering was to be delivered at the earliest possible date. It looks as if the Athenians were in need of time. They were probably already in the eighth prytany (1.60); the recommendation, given the force of law, contained a codicil stipulating that an extra Hekatombaion was to be added at the beginning of the following year.

Intercalation of a month was a normal remedy when the natural year was at an obvious variance with the festival year. But on this particular occasion the main purpose for the intercalation may have been to gain some time before the celebration of the Eleusinian Mysteries. This is suggested by the fact that normally it was the winter month, Posideon, that was doubled in Athens.[247] Posideon was well suited for this since it was the month when the solstice should occur. It was also, in contrast to Hekatombaion, a month poor in state festivals. We do not know if the insertion of an extra Hekatombaion meant that the Panathenaia would be celebrated twice.[248] In any case, a month's postponement of the Eleusinian Mysteries ensued, and on that occasion it was perhaps convenient to display the result of the year's harvest. The festival might also be the occasion for those who had to travel long distances to deliver their contribution.[249]

---

245 For a thorough discussion of the different views and argumentation for the 430s see Cavanaugh 1980, 33–100. Around 422 BC: Meiggs/Lewis 1969, on no. 73.425–22, 422/21 or 416/15 BC: Garnsey 1988a, 99.
246 *IG* I² 6. 91–2 and 122, cf. Cavanaugh 1980, 101–2, following the restoration of the inscription, ll. 32–3, by Clinton 1974, 10–12. *IG* I³ 6 has no *aparche* in its restorations, C.7–8 and 39.
247 Meiggs/Lewis 1969, no. 73; Bickerman 1980, 20; Samuel 1972, 58. Cavanaugh 1980, 117, has reservations concerning the possibility of pointing to a habitual intercalary month in the fifth century. The corresponding month was a normal intercalary month in Delphi, Aetolia; at Delos it was the twelfth month, Bickerman 1980, 20.
248 It is also impossible to tell if it was the year of a Greater Panathenaia.
249 Meiggs/Lewis 1969, no. 73. Some allies might have delivered the grain at the Panathenaia. From the year 425 they had to contribute a cow and a panoply to the Great Panathenaia, Meiggs/Lewis 1969, no. 69.96–7, *IG* I³ 71.56–7.

GODS AND AGRICULTURE

PURPOSE

The inscription also shows that so much grain was expected for Athens, or rather for Eleusis, that large containers had to be built for it. Contrary to the symbolic first-crop offerings of *eiresione* and *pyanopsion*, the sacrifice for Demeter and Persephone was to be put to use. The law makes it clear that *pelanos*, the best part of the grain, was to be used for sacrificial cakes in the course of the year.[250] The remainder was to be sold so that sacrificial animals and offerings could be purchased. Accounts from Eleusis show that, sometimes at least, part of the grain was sold and the money placed in the treasure-chests of the goddesses, divided between the one on the Acropolis and that which was found in Eleusis.[251] This provided a reserve capital for the goddess or goddesses, but if the decree took effect it also meant that a considerable amount of grain was offered for sale in or from Eleusis, grain that had not been grown in Attica.[252] It is perhaps an open question whether import tax was paid on grain of this type.

The amounts fixed by law concerning the first crop are not alarming and to the individual landowner cannot have been too overwhelming. We cannot see how much was left for the goddesses. The figures that have survived from the accounts of the fifth century are from war years, and we cannot tell which part of the amount indicated constitutes the total price of the grain. Cavanaugh estimates that there would probably remain sufficient amounts of grain for at least 10 talents during a normal year. In the fourth century the Athenians had no 'empire', but nevertheless, according to Isocrates, most states sent their *aparche* to the Athenians (*Panegyricus* 31).

From the year 329/8 we possess the figures for Attica itself and for the marginal areas such as Oropos, Salamis and Lemnos, 'the dependencies' as Garnsey calls them. During that one year grain delivered from these territories alone would suffice to meet the needs of approximately 1,666 persons for a year.[253]

How can we form an opinion as to the feelings of the landowners upon delivering the grain? It was the Athenians themselves who had imposed the dues, but perhaps their allies looked on it as a token of dependence on Athens when they were forced to export their offerings to Eleusis, and likewise when they were called upon to deliver an ox and a panoply for the Panathenaia. Isocrates, at any rate, writes that the first-crop offerings by

---

250 For the use of the *aparchai* see Chandor 1976, 186–8.
251 *IG* I² 313–14, covering the years 408/7 and 407/6 (*IG* I³ 386 = I² 313). The amount was the same both years and it was not very high perhaps because of the war. It is uncertain if the amount given was all that came in (Cavanaugh 1980, 206).
252 cf. *SIG*³ 976; *SEG* 1.366, about the Samians buying grain cultivated on the land of Hera on the coast of Asia Minor.
253 At the 'generous' consumption rate of 230 kg/person per year. There was 1.134 *medimnoi* of barley (×33.4) and more than 122 *medimnoi* of wheat (×40) – i.e., more than 383,180 kg of grain, according to the figures used by Garnsey 1988a, 104. Against abuse of these figures see p. 113.

non-Athenians were looked upon as a delivery to the Athenians; yet, he says, most states did it willingly as a token of gratitude because the Athenians had provided them with the grain. In other words, we are dealing with an offering consisting of a permanent, compulsory tax on production;[254] but the recipient is a deity, and the use of what is forthcoming from the first-crop offering is in the interest of the community: we are dealing with a divine service.

As we have mentioned, the term *aparche* does not by itself have any agricultural associations; for instance, it was also used to denote the sixtieth share that the Athenians paid to Athene out of the tributes given by their allies. Were Athene to have received her normal share, the term would have been *dekate* ('a tithe').

THE TITHE

The gods could receive a handful of the annual crop as *aparche*, or they could receive a fixed share of the crop in question. It seems that the share connected with the gods was often one-tenth, and that in fact they were to have a tenth of one's profit, be it war or peace. The expression is not used by Homer or by Hesiod; but according to Herodotus it was a common phenomenon, always referring to the share of the gods, but never, in fact, in an agricultural context. A Greek hetaira of Egypt, for example, offered one-tenth of her earnings to Apollo at Delphi, and with her money she purchased iron spits strong enough to carry an ox for him.[255] The god had the right to claim one-tenth of the spoils of war.[256] It is in this connection that we find the expression used by Xenophon.[257] When Cyrus' army in Asia Minor was dissolved, Xenophon and his fellow general took it upon themselves to see to it that a *dekate* ('a tithe of the booty') would be secured for Apollo and Artemis. Apollo, straightaway, received a present as his share of the booty, but it was not until after some delay, owing to the conditions of war, that Xenophon bought a piece of land for the half of the tithe that he was supposed to administer on behalf of Artemis; this piece of land, to please the goddess, was somewhere in Skillus. Each year he took a tithe off the crop of the land and used it to uphold the cult by celebrating a great festival, and he set up a stele at the site of the sanctuary with the following inscription:

THIS AREA IS CONSECRATED TO ARTEMIS: WHOSOEVER HAS IT IN HIS

---

254 *Aparche*, termed *telos* in *IG* I³ 130 of c. 432. cf. Meiggs/Lewis 1969, no. 89; *IG* I³ 101, concerning Parthenos in Kavalla. The permanency can be seen also from *IG* II/III² 140.14 of c. 353/2.
255 Herodotus 2.135. Note also 3.55, merchants from Samos set up a krater in the Heraion as a *dekate* of their gains at Tartessos.
256 Herodotus on *dekate* from the spoils of war – to Zeus (1.89); to Athene (5.77); to Apollo (7.132); to Delphi (9.81).
257 The term *aparche*, on the other hand, is all in all found only twice in Xenophon, *Oeconomicus* 5.10 and *Hieron* 4.2.

POSSESSION AND MAKES USE OF IT, MUST OFFER A TITHE EVERY YEAR. FROM THE PROFIT, HE WILL MAINTAIN THE SANCTUARY IN ITS PROPER CONDITION. SHOULD ANYONE NEGLECT THIS, IT WILL NOT PASS UNNOTICED BY THE GODDESS.[258]

In other words, here we find an example of the tithe: a tithe of the spoils of war is converted into sacred ground, which becomes the property of Artemis, and from this, in its turn, a tithe is taken to maintain the cult. It is clearly seen that it is to be taken quite seriously, with regard to the question of ownership, when we read, 'The Goddess provided the guests with flour of barley, loaves of wheat, wine and dried fruits, a portion of the sacrificial animals from the holy pasture, and a portion of the game' (5.3.9).

BLOOD SACRIFICE

In Eleusis part of the offerings of the first crop in the form of grain from many parts of the Greek world was to be converted into sacrificial animals for the benefit of the goddesses. In Xenophon's Skillus something rather similar took place. The goddess issued invitations to a festival for the tithe that her own land had yielded for the particular year, and at the festival she served meat. Xenophon views the situation from the materialistic point of view with a focus on the meal. In the decree concerning the first crop to Eleusis, on the other hand, the emphasis is laid on the fact that we are dealing with an offering. It is, however, the same issue: the offering of the first crop is not identical with the offering of an animal, but it does in fact, in these two cases, also include offerings of animals. When the decree speaks of offering the first crop (*aparche*) the motive is indicated as well as the circumstance that what is offered is part of a whole, a (god-given) profit; when it deals with offerings of animals, often called *thysia*, all that has been said is that ritual killing is concerned, not what the occasion is nor how the animals to be sacrificed are paid for.

There are many approaches to the study of the sacrifice of animals; but in a book concerning agriculture the essential point must be which, and how many, animals were offered, from where the animals came, and what was done with the meat.

Sacrificing of animals is met with on all levels of Greek society. It could be the city-state as such that performed the offering. No popular assembly in Athens could be introduced without a sucking pig having been offered. At the Panathenaia, Athena was to have a complete hecatomb (100 oxen);[259] but similar and even larger offerings are known from other Greek areas.

---

258 Xenophon, *Anabasis* 5.3.4–13. For discussions of this passage see Skydsgaard 1988b, 85 note 27, and with a diverging interpretation Hodkinson 1988, 48.
259 In 410/9, a year of war, the Athenians used 5114 drachmas for the hecatomb at the Great Panathenaia, Meiggs/Lewis 1969, no. 84.6–7.

AGRICULTURAL PRODUCTS FOR THE GODS

According to Xenophon, in the year 370, Jason ordered the *poleis* in Thessaly under his command to prepare oxen, sheep, goats and pigs for offering at the imminent Pythia, that is to say the games at Delphi. Not every city was called upon to supply so many, but nevertheless the upshot was more than 1,000 oxen and more than 10,000 of the other animals. The animals to be sacrificed had a long way to walk. It must have been an impressive procession, manifestation of power and a colourful spectacle at Delphi.[260] Jason cannot have been the only one bringing animals for sacrifice and Xenophon's account seems almost unbelievable when we consider that so many animals were to be fed on the way (see p. 105). Another manifestation of power is found in Syracuse when, in the third century, Hieron II erected the greatest known Greek altar, approximately 200 m in length. Perhaps it was made that size so as to be able to carry the 450 bulls that were to be sacrificed annually to Zeus Eleutherios (Diodorus Siculus, 11.72.2; 16.83.2).

Private people individually offered a sucking pig in connection with initiation into the Eleusinian Mysteries. This was a prerequisite for participation. Were they to seek help at some religious sanctuary, it was possible that offering of an animal to the particular deity would be required. In complete privacy, slaughtering of an animal was arranged at home as an offering; in fact, every slaughter was a sacrifice.[261]

SCOPE AND DESCRIPTION

Information concerning the rhythm and the scope of private offerings of animals leaves much to be desired, but we can safely say that it varied in accordance with the status of the person and the local conditions for cattle-breeding and hunting. In an area like Attica offering of animals at home was a rare phenomenon for the greater part of the inhabitants. Information concerning the consumption of sacrificial animals on the part of the Attic state could perhaps be gleaned, in very general terms, by combining our knowledge of offerings like those that, as a matter of routine, were regularly performed before a meeting with our sporadic knowledge of what was sacrificed in connection with state festivals. However, the picture would inevitably become vague and distorted.

With regard to the smaller entities in the city-state of the Athenians, the demes, fragments of calendars which refer to offerings have been found, particularly during the past twenty-five years. Only one of them is almost complete, the one applying to the deme known as Erchia, a deme east of the Hymettos Mountains.[262] But fragments from four other demes enable us to

---

260 Xenophon, *Hellenica* 6.4.29. Jameson 1988, 95.
261 For a famous description of an offering in comedy see Aristophanes, *Peace* 925–1126.
262 Fragments of the Athenian sacrificial calendar have been found, see Dow 1968.

Month	Piglets	Sheep Adults		Lambs		Goats Adults		Kids		TOTAL
		M	F	M	F	M	F	M	F	
Hekatombaion	2					2				4
Metageitnion	2	2	3	1		1				9
Boedromion	2	2	3	1						8
Pyanopsion			1							1
Maimakterion										
Posideon	1		1							2
Gamelion	2		4	1	1	3	1			12
Anthesterion			1						1	2
Elaphebolion						1	1			2
Mounychion			3							3
Thargelion			5		1	1	1			8
Skirophorion	1		2	3						6
TOTAL	10	20	11	1	3	5	6	1		57

Note: M = male; F = female.

Figure 12.1 The offering of animals in individual months, Erchia

put the Erchia calendar into perspective. It is only natural that initially the deme calendars gave rise to a series of important articles.[263]

The great advantage of the deme calendars is the fact that they are handed down direct, on stone; in addition, they have a direct bearing on the subject of the present chapter, namely the offering of animals. We shall have a closer look at the Erchia calendar because it is preserved almost completely and may be regarded as typical of a fairly large, but not especially wealthy, Attic deme (Figure 12.1).

In the deme calendar, like the official Attic year, the year starts in the middle of summer with the month Hekatombaion. In five columns on the stele whenever an offering is to take place there is an indication: *date – deity – place – sacrificial animal – price*. In the chart we have produced, we have not separated the individual days of offering but solely considered which and how many animals were to be used in the particular month. In the Erchia deme no hecatombs were offered; in fact, not even a single ox was offered. But when the offering of a sheep to Athene was performed, the animal was called an *antibous* (an 'instead-of-an-ox-' sheep), so it must be for reasons of economy that offerings of oxen are lacking (col. 1, 62–5). In Marathon there were fewer, but more costly, offerings, and here the offering of at least six oxen in the course of a year was prescribed. The absence of horses in the Erchia calendar testifies to something that is valid generally. In

---

263 Mikalson 1977; Daux 1963 and 1983; Dow 1965; Parker 1987; for a nuanced discussion of our theme see Jameson 1965 and 1988; Whitehead 1986, 185–208.

the Mycenaean period, the Greeks did allow a dead man's horse to be buried with him in the grave, but otherwise they did not sacrifice horses to the gods. Nor is game mentioned in the Attic festival calendars, most likely because you could not be sure that the game would be available at the right time.

The calendar is of course normative in character. The prices of the victims are ideal prices and the sex of the victims can be deduced from the prices prescribed. Female victims were cheaper than male, probably because they were smaller. In Erchia the deities according to the calendar could expect victims of their own sex.[264] An exception must be made for the piglets. The price of a piglet was 3 drachmas irrespective of the sex of the recipient deity. It is doubtful that this means that all the piglets sacrificed were of the same sex. Rather, the piglets were considered sexless while still immature. Also there is no difference in the quality of the meat at that stage. As we can see, the sheep is the most common sacrificial animal in Erchia; next comes the goat, and last the piglet.

In the Erchia calendar, beside the usual formula, it is also often stated what should be done with the meat. Upon one rare occasion the entire sacrificial animal is to be burned. In that case, the term is a *holocaustal* offering, and in connection with such offerings it is indicated that no wine is to be consumed. Normally no reference is made to wine, which must indicate that wine was required, and as a rule the idea was that the animal was to be eaten.

APPLICATION – 'LE PARTAGE'

In recent years the essence of animal sacrifices has been studied, particularly in France among historians of antiquity or of religion, inspired by anthropological research. The core of the discussion is felt already in the title of the article by Nicole Loreaux, 'La cité comme cuisine et comme partage' (1981c).[265] Roughly speaking, Xenophon's point of view permeates the French treatment of *thysia* (the blood sacrifice) – the offering of an animal is a feast. Those who participate in the meal thereby emphasize their feeling of alliance with the god to whom the offering is given, but also, particularly, their mutal alliance. Thus those who do not participate are defined as outsiders.

On the purely private level, the mechanism is seen in a speech concerning hereditary rights made by Isaeus where the speaker needs to prove his legitimacy. Before the judges he has to prove that his maternal grandmother was legally married to Chiron, the testator. He claims that the late Chiron is his grandfather, and to prove it he relates the following:

---

264 This was a tendency but not a rule in Greek sacrificial practice, Dow 1965, 187.
265 A critical commentary on Détienne/Vernant 1979.

as was natural, seeing that we were the sons of his own daughter, Chiron never offered a sacrifice without our presence; whether he was performing a great or a small sacrifice, we were always there and took part in the ceremony . . . and when he sacrificed to Zeus Ctesios – a festival to which he attached a special importance, to which he admitted neither slaves nor free men outside his own family, at which he personally performed all the rites – we participated in this celebration and laid our hands with his upon the victims and placed our offerings side by side with his.

(Isaeus 8.15–16)[266]

In the city-state it was the popular assembly that decided who was to have a share of the meat offered, and according to which rules. As for the hecatomb for Athene in connection with the smaller Panathenaia, for instance, in a decree from the fourth century (IG II^2 334.21–7) we read:

When they [the *hieropoioi*] have sacrificed to
Athene Polias and to Athene Nike
all the cows they have bought with the 41 minae
they shall distribute the meat to the Athenian people
at the Kerameikos, just like at other distributions of meat.
And they shall apportion each deme its parts
according to the number of participants it renders for the
procession

By regulating the participation of the offering, and by distributing the meat to 'the Athenian people' – that is, to the citizens in relation to their participation – it is indicated who are members of the citizenry. You are joined together in the common meal consisting of the precious beef.

Admittedly, in the Erchia deme no one could afford beef, but in about half of the cases the price of the sacrificial animal was followed by the decree *ou phora* ('not to be carried away'). In other words, the meat was to be consumed by the participants of the offering on the spot.[267] In a couple of places the phrase 'should be surrendered to the women' is added. In these cases it was a women's festival where, perhaps, the only male participant was the man who performed the slaughtering, and where the meat was eaten by the women at the festival.

A similar decree concerning the consumption of the sacrificial animal in the place where it was offered is known from the oracle and the divine sanctuary, the Amphiareion, in Oropos. Apart from the entrance fee, the individual patient would have to sacrifice an animal, and meat from the latter was not to be taken from it. In this way measures were taken that the right

---

266 Translated by E.S. Forster, Loeb edn.
267 cf. Dow 1965, 208–10.

AGRICULTURAL PRODUCTS FOR THE GODS

circle received its share of the meat at the right time; in other words, the meat was not to be sold on, nor salted for later use.[268]

The god himself, not fond of human food, received as his share of the offering the smell of fat emanating from the burning of less-digestible parts of the meat. In Oropos the priest could claim a bone from the animal whether it be the wing of a bird or the leg of the sheep. At the smaller Panathenaia, the rules for certain lesser offerings ran as follows:

> and they [the *hieropoioi*] shall apportion
> the prytanes five parts and the nine archons
> three and the treasurers of the Goddess one
> and the *hieropoioi* one and the generals and the taxiarchs
> three and the Athenian participants in the procession and
> the basket carriers in the usual proportion, and the rest of the
> meat they shall thereupon distribute among the Athenians.
>
> (*IG* II² 334.10–16)

The inscription attempted to secure privileges of this kind, but there was also in Erchia an attempt to safeguard the less-privileged against abuse from those who were customarily better off when it was committed to stone that meat must not be removed from the place of offering. Here, it was also felt that it would be necessary to write 'the hide to the priestess', when she was to have it; but we never hear about 'the hide to the priest', though it is almost certain that it was the priest who would automatically receive it unless something else is specifically indicated on the stone. At the Panathenaia the hides from the 100 sacrificed oxen were sold. The profit went to Athene, and in 334/3 it amounted to more than 10,000 drachmas.[269]

FINANCING AND DELIVERY

The custom of sacrificing animals is older than the era of the city-state, and as for many of the regular offerings the responsibility for them, and financing them, involved a slow process with sacrifices being taken over from private families and becoming official. It never became perfect. The Erchia calendar, as seen by Dow, testifies to the effect that families who from time immemorial had financed the offerings, could no longer manage them. So the burden from then on would have to be distributed among the wealthiest in the deme according to the principle of liturgy. The odd distribution on the stone, into five columns of offerings that cost about the same, probably means that each year the deme appointed five members who in that particular year were responsible for the offerings in one of the five columns of the stele. Frequently, however, the offerings of animals were financed by the

---

268 cf. Theophrastus, *Characters* 9.1–3 about *anaischyntia*.
269 *IG* II/III² 1496, cf. Jameson 1988, 96, with a cautious attempt to calculate the number of sacrificial animals involved.

gods themselves, such as when, for instance, sacrificial animals were purchased for the first-crop offering in Eleusis. Where the animals came from has been discussed in Chapter 5. So modest a number of sacrificial animals as is mentioned in Erchia could no doubt be delivered by the liturgists themselves, or they could easily be purchased in the deme. To some extent the calendar of the offerings of animals reflects the rhythm showing which animals could most easily be dispensed with at the various times of the year.[270]

In certain cases the gods themselves did in fact own sacred animals among which the animals to be offered were chosen. For instance, this occurred in Skillus with Artemis, and at Delphi with Apollo, and on Delos. These sacred animals were grazing on the land owned by the gods themselves.

---

[270] Jameson 1988, 102–3, with some reservations.

13

LAND BELONGING TO THE GODS

It is not self-evident that land belonging to the gods should be treated as a category by itself as we have chosen to do in this book. In 1952 Moses Finley lent his name to the well-known and by now universally accepted dogma that, among the Greeks, the important dividing-line as far as land ownership was concerned separated public land owned by the state from the remainder, whereas it was of no consequence, legally or otherwise, whether land was administered via temple funds or not.[271] In a recent article Robin Osborne treats the leasing of land and buildings in Classical and Hellenistic Greece under one heading.[272] Unlike Finley, who distinguishes, first, public (to be understood as 'land owned by the state') and, second, all other land, Osborne sets up two categories: first, public and corporate property, sometimes merely called 'public', and, second, private, as an equivalent to land 'individually owned'. Occasionally, Osborne will refer to a piece of land as 'sacred' or 'templeland', always listed under his first category.[273] Here we are dealing with a practical, but not entirely closely reasoned, dividing-line into categories. Osborne agrees with Finley in finding no cogent reason why land owned by a god should be regarded as a separate category; but we, on the other hand, shall draw our conclusion from the fact that the Greeks themselves were fully aware when the question concerned land belonging to a god.[274] This we shall do because it has turned out to be of great practical importance in agricultural contexts.[275]

---

[271] Finley 1952, 95. cf Whitehead 1986, 170.
[272] Osborne 1988. The Classical period is represented by Athens, the Hellenistic by Delos, Thespiai and Karthaia.
[273] It would be difficult to place the land of Artemis in Skillus in this classification, see p. 173.
[274] See e.g. the terminology used in Aristotle, *Ath. Pol.* 47.3–4.
[275] cf. Parker 1983, 160–6.

GODS AND AGRICULTURE

COMMON PREREQUISITES

A piece of land belonging to the gods was known as a *temenos*, a piece of land set aside.[276] Such land was also described as *hiera* ('holy' or 'sacred'). Whilst we have seen that considerable differences existed within the individual city-state as well as between various city-states where land owned by human beings was concerned, there is good reason to expect that a higher degree of unity and stability would apply when land belonging to gods was involved. Here, certain common characterictic features may be observed:

1 the owner, be it god or goddess, never made his or her appearance;
2 the owner, therefore, never cultivated the land in person;
3 the owner gave no directions of any kind;
4 the owner never died, so that there would never be a question of distribution of the land by inheritance; and
5 the owner did not sell the land.[277]

In other words, the responsibility for the administration of the land owned by a god rested upon the people who felt that they depended on the favour of that particular god. The fact that the owner was always absent, and that the administration of the land owned by the god concerned a wider circle means that today we possess a rich epigraphic source of material that throws light upon the general lines of the problem as well as accounts in individual cases. The circumstance that change of ownership never occurred means that the risk of attempting to make general statements whenever land owned by a deity is concerned is less than it is when land owned by human beings is concerned.

THE ACQUISITION OF LAND BY THE GODS

As we have seen in the section on the status of land, common practice required that land was reserved for the gods whenever a new city was founded.[278] In his *Laws* Plato has his Athenian express explicitly that no person in his right mind, on an occasion like that, would omit, initially, to present to the gods exquisite or special (*exaireta*) plots of land and allocate to them everything that was their proper due. Here, the Athenian refers to a practice that he regards as compulsory to comply with in connection with the founding of the ideal state (Plato, *Laws* 738c–d).

In the cities that had not, in the true sense of the word, been 'founded', but rather developed into self-grown city-states, the gods were also, to a

---

276 The ancient etymology from *temno* is questioned by several scholars, Hegyi 1976, 78, Malkin 1987, 140, note 24, cf. Frisk 1960, s.v.
277 The so-called *rationes centesimarum* might reflect one of the exceptions, bound to be found; but see the tempting suggestion that they deal not with sale but with leasing, Osborne 1985a, 56–9.
278 Malkin 1987, 138–42.

greater or lesser degree, landowners; but for obvious reasons their pieces of land were not placed according to a system that had been established once and for all. Yet, as a rule, land owned by gods would be found along the borderlines of the city-state. Such land had its own particular function, a subject to which we shall return.

Sometimes the gods could acquire new land. It could be awarded them by the popular assembly as their share of newly conquered land,[279] or the officer in command could purchase an area for them 'at home' as their tithe of the booty, as Xenophon did for Artemis in Skillus.

Delos had always been a sacred island, the implied reason being that Apollo and Artemis were born there. Even if the island as such was sacred, it did not follow that Apollo possessed all land on the island. In 525 Polykrates, the tyrant, conquered the nearby and larger island of Rheneia and presented the island to Apollo.[280] The Rheneians were displaced to a corner of the island so that it seemed as if it were the people who had had part of the area set aside for themselves from the land of the gods, but terminologically the fiction was upheld that the portions of land belonging to the god were referred to as *temene* ('lots cut off'). In 417 Apollo's property on Delos was extended when Nikias, the Athenian general, from his own private means bought part of the island at the cost of 10,000 drachmas, and presented it to the god.[281] A foreigner would scarcely be able to have done this in Attica, the native state of Nikias, and it is doubtful whether an Attic citizen would have been able to do it. The Athenians had strict rules governing ownership of land as well as political power to maintain such rules. They could have no interest in witnessing that too much land should be withheld from normal rules applying to ownership and inheritance, which was exactly what happened when land was given to the god.[282] *Horoi*, which marked security for a dowry, is an almost entirely Attic phenomenon. A few *horoi* from Amorgos have been preserved, and it is probably not by coincidence that it is an Amorgian and not one of the many well-attested Attic *horoi* that shows that the property placed as security for the wife's dowry, in case of her death or the death of husband and wife, should be consecrated to Aphrodite, thereby passing from private property to property belonging to a deity.[283]

In the registers of leasing of property belonging to Athene and the other gods, when either a house or a piece of land is described as 'that which Kallikrates sanctified',[284] it concerns Salamis and not Attica itself. It is

---

279 Thucydides 3.50.2 on Lesbos.
280 Thucydides 1.13.6, 3.104.2. Rheneia is 16 km², Delos only 6 km², Cavagnola 1973, 512.
281 Plutarch, *Nicias* 3.6, cf. Kent 1948, 256; Hands 1968, 57–8.
282 It was probably possible to sell land immediately, if it came by accident into the hands of the god, e.g. as a result of confiscation, cf. Guiraud 1893, 376.
283 Finley 1973c, no. 155.
284 Stele I col. iii C. 6, Walbank 1983, 108. The part of the inscription that indicated the kind of property concerned has not been preserved.

probable that the verdict laid down by the Athenians to the effect that officials, out of consideration for the imminent presentation of accounts, were not allowed to dedicate their property to the gods, shows that private persons *could* dedicate their land to the gods also in Attica, although the expression of property is *ousia* and thus not necessarily having to do with land (Aischines 3.21). The circumstance that the Athenians laid down limits concerning the right to sanctification of land may also appear from the decree relating to the colonization of Brea. The surveyors employed by the colonists were there under order to respect the sacred areas that had already been defined as such in Brea, but not to add new *temene*. Perhaps this means that the surveyors were to respect what was already sacred. More specifically it may indicate that the part of the procedure of founding that related to setting land aside for the gods had already taken place.[285] In *Religion and Colonization in Ancient Greece* (1987), Irad Malkin suggests that this is the second Athenian attempt to colonize the place, and that the decree stipulates for the sacred areas set aside during the first attempt to be respected.[286] Whichever solution may be the right one, it probably suggests that at any rate the sacred area was not to be extended.

THE UTILIZATION OF SACRED LAND

Plato gives us a clear explanation of why it was so important for the gods to receive good *temene* in each section of the colonies:

> so that, when assemblies of each of the sections take place at the appointed times, they may provide an ample supply of things requisite, and the people may fraternize with one another at the sacrifices and gain knowledge and intimacy, since nothing is of more benefit to the State than this mutual acquaintance.
>
> (Plato, *Laws* 738d–e)[287]

In other words, the regular festivals which bound the members of the state together were to be financed from their *temene*. This was also the intention with the land that Xenophon purchased for Artemis. Perhaps 'financing' is not the right term. In both cases the idea was that you were to eat and drink from what the land offered; so the land had to be cultivated.

In the area allotted to Artemis, cattle were also to be raised, and there was game which could be hunted for the festival.[288] No doubt this meant that her territory in Skillos contained a mountainous area as well as the arable plain. The place where Skillos was located (as this is assumed today) has all these

---

285 Thus Meiggs/Lewis 1969, on no. 49.9–11.
286 Malkin 1987, 155–60.
287 Translated by R.G. Bury, Loeb edn.
288 See p. 174.

prerequisites (there is also a very meadowy area just as Artemis would like to have).

Mountain ranges frequently constituted borders between the city-states. Here, the gods often owned land which could be utilized for grazing and for hunting. In a good survey of the problems concerning frontiers, Sartre rightly emphasizes the contrast between, on one hand, the border area of a city-state, the wild, the domain of Artemis and Apollo, the world of the ephebes, and on the other, the well-organized world, that of cultivated fields and the city, the adult world.[289]

At times, the borderline between two city-states ran through arable land. Such land might be owned by a deity, and it could be specifically prohibited to till this soil. A well-known example is the Holy Orgas,[290] land which lay on the border between Megara and Athens or, perhaps, more correctly on the Athenian side of the frontier. In any case, the land belonged to Demeter in Eleusis and could not be cultivated, although it was fertile. The area ended up as a 'no man's land' under divine surveillance, a buffer zone.

THE ORACLE AND THE LAND OF THE GODS

Whenever questions of doubt arose concerning land owned by a god, the right thing to do was to approach an oracle, often the Oracle of Delphi. When a colony was to be established, a confirmation was requested, usually from Apollo, concerning the plans entertained. What an oracle had pronounced concerning sacred land was not to be contested – so Plato as well as others claimed (*Laws* 738b–c). Xenophon, too, obtained confirmation at Delphi with regard to his plans concerning land for Artemis.

An inscription from Athens from the year 352/1 (*IG* II² 204) shows how the Holy Orgas, without any interference, had gradually diminished. The boundary-posts had been overturned or moved and private persons had slowly taken more and more of the land for cultivation. At the popular assembly it was now decided that sessions were to be held several days in order to arrive at a settlement of the border conflicts. In the future the Areopagus Council, the *strategos* in charge of the *chora*, peripolarchs and all Athenians were to make sure that the boundary-posts remained where they were supposed to be. Messages were then to be sent to Apollo at Delphi in order to obtain an answer to the question whether it was better to leave the Holy Orgas untilled, or whether it was better to have the area cultivated and use the money for building on the sanctuary of the goddesses in Eleusis. Apollo's answer is unknown, but the alternative that was offered to Apollo instead of leaving the land untilled was leasing-out. There were no other acceptable solutions.

---

289 Sartre 1979. See also Jameson 1989.
290 For the significance of *orgas* see Pritchett 1956, 267–8.

In the sixth century, at Delphi, Apollo was confronted with the question of what was to happen with the city-state of Kirrha which had caused a holy war but had been defeated. Apollo's answer was that the city of the Kirrhaians was to be demolished, the inhabitants expelled, and that their territory should forever remain fallow (Aischines 3.108–12).

After the Greeks had been defeated by Philip of Macedonia in 338, the king of the Macedonians presented the complaisant Athenians with the small city-state of Oropos. There are five mountains in the area; they were divided between the *phylai* after lots had been drawn and after the *phylai* had been paired so that two shared each mountain. After a while it became apparent that the mountain that the *phylai* Akamantis and Hippothoontis had received, at some previous survey or adjustment by horists had been marked as belonging to the god Amphiaraos. In order to obtain a manifestation from the god in this embarrassing question, the popular assembly appointed three men, among them one by the name of Euxenippos. They were to go to the sanctuary of Amphiaraos at Oropos and spend the night there in order that the Athenians might learn what Amphiaraos would let them dream. Among other things, this particular sanctuary was also an oracle of dreams. Euxenippos returned with an ambiguous answer which enabled the Athenians to interpret it in such a way that they would not have to change their plans with regard to the controversial mountain. But an orator named Polyeuktos now suggested a decree that the two *phylai* concerned should return the mountain to the god whilst the other eight *phylai* would then indemnify the two. The proposal was voted down as being incompatible with the dream. Now, our informant feels, the right thing would have been to send a messenger to the Delphi Oracle in order to obtain an interpretation. Instead, Polyeuktos denounced Euxenippos for subversive activity, and for the use of one of those who supported Euxenippos, Hyperides wrote the speech (3) that is our source on the whole matter. The example shows how the god could be consulted in cases of doubt. Here again one sees how the god could not be deprived of the land allotted to him or her, no matter how much time had passed and how political conditions might have changed. In this connection it did not matter whether the sacred land constituted a specific legal category.

ADMINISTRATION OF THE LAND OF THE GODS

The leading principle was, then, that land which had come into the possession of a deity should remain so. The mortal administrators of such land could change; administrators in this connection are to be understood as the city-state, the corporation or the individual person who had, or took upon himself, the responsibility for the cult of the particular god in the individual place and not the persons who were responsible for the daily running. The mechanisms in connection with the administration of the land of gods become particularly clear in an example like Delos which was so sacred to all Greeks that from the year 424 no one was allowed to give birth or to die on the island.[291] Beyond dispute, Apollo was the greatest landowner here and on the neighbouring island of Rheneia.

His *temene* were administered, as was the god's other property, by the people who had political power and will to do so at any given time, or by those who were entrusted to do so by such persons of influence. From 478 to 314, with a brief interval from 404 to 394, the Athenians were the administrators. From 314 to 166 the administration was performed by the Delians themselves because more powerful agencies allowed them to be independent. Thereafter the Romans gave the island to the Athenians, from which time it was no longer merely the sacred land, but the entire island that was concerned. The Delians were expelled, and the Athenians decided who would now be allowed to utilize the land, the sacred land as well as that owned by private persons. It is interesting that during the long history of Delos the changing administrators did not alter the way in which the god's land was managed. They were confident that it was to be leased out, and the individual farms or lots continued to be known, in the accounts, by the names they had always had, and required no closer description.

Within the territory of the city-state of Attica, Attic demes were responsible for certain sacred lots of land in their area. Apart from that, administration was controlled from Athens.[292] The important and constant feature was that the lots of land were to be leased out and the income thus derived to be used for the maintenance of the cult of the particular god or goddess.[293] Formerly, the establishment of the board of officials, 'The Treasurers of the Other Gods', during the later half of the fifth century was interpreted as a general secularization of the Greek society. This interpretation has been rejected clearly by Tullia Linders who points out that even the part of the gods' income paid into the fund and administered by the Treasurers continued to be listed and treated as belonging to the god who had received the income.[294]

In this way the income of the gods differed from state income which, in Athens, was not in principle ear-marked for a specific purpose. It was a side-effect that the city-state or deme or perhaps the private person who administered the income that the gods derived from the land might thereby find it possible to manage a surplus on behalf of the god. The surplus could be considerable and permit the administrators in question to have the gods

---

291 Thucydides 3.104; *IG* XI 2 145.8–9. This is normal on sacred land, according to Parker 1983, 161–3. The exceptional thing is that the prohibition covers the whole island of Delos.
292 The Assembly was responsible for the Hiera Orgas which belonged to the goddess in Eleusis, see p. 185. We meet centrally administered leasing-out of god-owned property in the leasing-lists discussed by Walbank, 1983, 1984, 1985.
293 At Delos the income was also used for the cult of the other gods at the place.
294 Linders 1975, 67.

grant loans from it. In many cases, however, the income was not sufficient to maintain the cult, with the result that, instead, the administrators were faced with the problem how to cover the deficit.[295]

Until recently it was thought that the law concerning Nea (see p. 141) testified to an unusual initiative owing to difficulties in financing the Panathenaia. In the law it is specified that the income from leasing the area in question was to be spent for the purchase of the hecatomb for Athene during the Panathenaia. It was looked upon as something out of the ordinary, but also as a sign of the times, that regular state income was ear-marked for one particular purpose in that way. But if Nea was an island dedicated to Athene, as suggested by Langdon, the example is rather more reminiscent of Delos, although the goddess for whom the money was meant in this case resided far from her island, on the Acropolis of Athens.[296] If so, it still shows the difficulties connected with the financing of the cult, but the method adopted was perhaps not so unusual. It may be argued against this interpretation that the *poletai*, not the King Archon, managed the leasing. As a rule the latter managed the leasing of the gods' land.

LEASING THE PROPERTY OF THE GODS

In the ideal states of the philosophers tenants were not required, not even for the sacred land, which was thought of as being cultivated by dependent labour. Probably the same applies to the city-states of the time that inspired the philosophers, and had dependent farmers at their disposal – such as Sparta and the city-states on Crete. Normally, the city-states had no such dependent labour at their disposal to any great degree. Therefore the administrators of the sacred land had to find tenants. At the public assembly in Athens, for instance, general lines were laid down regarding the leasing of divine land administered by the state, whether in Attica or on Delos. In the assembly of the demes, similar procedures took place concerning sacred land administered by the deme.[297]

The size of the plots of land

It is rarely stated in an inscription concerning leasing how large the particular piece of land is.[298] Often the lot has a name from former days, and the readers of the time required no specific description: this, for instance, applies to Apollo's lots of land on Delos and the *temene* in Piraeus administered by the demes. In other cases the boundaries of the lot are described, but as we do not know them today, this is of no help to our knowledge of the area.

---

295 One of the main reasons for having a budget in the deme were the deme's cult obligations, Whitehead 1986, 164.
296 *IG* II/III² 334; Lewis 1959b; Langdon 1987.
297 See especially *IG* II/III² 2498 = Behrend 1970, no. 29 = Isager 1983, no. 2.
298 Exceptions are *IJG* 12 and 13.

In rare cases we know that the tenant had income from elsewhere, apart from the lot of land leased. This does not mean by itself that he could not have lived solely from the land leased. We can only arrive at an impression of the size if the inscription contains stipulations as to what is to be grown or planted. Some information may also be gleaned from the size of the rent, but this is dangerous if we know nothing about the qualities of the particular piece of land. Nor can we know whether the size of the rent was artificially modified or increased. The former could be the case if the demes used the leasing as some sort of dole, or indeed as a reward, but one might also imagine that a certain amount of prestige was connected with leasing the sacred lots of land at an excess price.[299]

Since rents of sacred land in Hellenistic Thespiai were often very low, it has been suggested by some that here sometimes very small lots of land were involved, or that lots of limited value were leased out.[300] Their attraction may then have been their location.[301]

It seems likely, however, that farmland belonging to gods was as a rule leased out as complete farms or as units from which you could make a living and even have a surplus to pay the rent.[302] This must have been the normal situation when a new colony was founded and probably also when land was testated to a god. Clearly, it applied to Apollo's land on Rheneia which was leased out as ten farms, each consisting of undivided land; thereby they differed from private estates which were often divided by distribution to heirs.

The identity of the tenants

Whereas ownership of land normally presupposed citizenship where the land was located, we know of no decrees to the effect that a tenant must belong to some definite social class. It is not likely that such laws normally existed, but this does not mean that those who administered the leasing of the land of the gods may not have had an implicit agreement to that effect. On a few occasions in the Attic inscriptions on leasing of gods' property, metics are mentioned as tenants, but only in one case is it definitely a question of a piece of land.[303]

In Piraeus everyone other than citizens was in practice excluded from bidding for leasing of land belonging to gods because real property was required as security. One might think that it was actually with this in mind

---

299 Whitehead 1986, 156–8, with references.
300 Osborne 1988, 194–7.
301 cf. pseudo-Aristotle, *Oeconomica* 1346a 13–26 concerning Byzantion. This passage has not as yet found a convincing interpretation.
302 Osborne 1988, 292–7, argues for a different pattern in Thespiai.
303 Stele I col. iii C.5 (a house); stele I col. i A.5 (character of property unknown); stele I col. iii C.1, Walbank 1983.

that the rule was introduced, inasmuch as the Piraeans lived in a heavily populated area with an exceptionally high number of metics.[304]

Osborne claims that leasing of public and corporate property was in great demand everywhere, thus including what we call land owned by gods, and that it was difficult, if not impossible, for people of modest means to enter the picture. As for Attica, it may seem that those who administered land on behalf of a god would have a tendency to choose a tenant from among themselves: the state a citizen, the deme a member of a deme, the *orgeon* a member of the *orgeones*.[305] However, there is no rule without exceptions,[306] and the material is altogether too flimsy to allow any conclusion as to how often that pattern was followed. Choosing someone from one's own circle was not without problems, as is clearly demonstrated in Demosthenes' speech *Against Euergos*. The man for whom the speech is written had been voted out of the list of citizens by the members of his deme. He carried the case to the people's court in Athens at the risk of being sold off as a slave, and now, in the speech, he explains his precarious situation by stating that he has made enemies among members of his deme. As a *demarchos*, he had seen it as his duty to collect debts owed by certain members of the deme, debts consisting primarily of unpaid rents for sacred land (Demosthenes 57.63).

On Delos, too, tenants of Apollo's land were mostly wealthy men, at least during the period of independence. From before that time, we know hardly any names of tenants.[307]

---

304 *IG* II/III² 2498.3–5. Isager 1983, 32–3; Whitehead 1986, 157; Osborne 1988, 289 note 28.
305 Osborne 1988, 290–1.
306 The theatre in Piraeus was leased to two men from the deme and two from outside, Behrend 1970, no. 30, Isager 1983, no. 3.
307 Kent 1948, 320.

14

THE ANIMALS OF THE GODS

Not only could the gods own land; they could also own animals. This fact is beyond a doubt, although it was probably less common for gods to own animals than to own land; in any case the evidence is much more scarce in our sources than that which applies to gods owning land. The sources are almost entirely epigraphical.

One of the inscriptions concerning animals belonging to gods that is most frequently quoted is the law concerning the sanctuary of Zeus on Amorgos,[308] which lays down the rules for the leasing of land belonging to Zeus. One of the decrees for leasing is that the tenant is not allowed to have cattle grazing on that land. Were he to transgress this law, the cattle in question would be regarded as sacred, that is, the property of Zeus. Whether they could then continue to graze as Zeus' cattle, or whether they would soon be sold or sacrificed, the inscription does not show. It does not immediately appear as if grazing on Zeus' land was anticipated. That, on the other hand, was the case with a sacred area mentioned in an inscription from Ios on the border between Laconia and Arcadia. Here the decree was that anyone who pastured animals, *thremmata*, on the land of the particular god was to brand them so that it could be seen that they belonged to the god.[309] The branding shows that this was not merely a question of transition, but of a more permanent grazing.

Cattle belonging to gods are also mentioned in connection with a piece of land belonging to Alea, or Alea Athene, in Tegea.[310] The sanctuary was located on a plateau, and Athene's land was well suited for grazing. Access to this privilege was regulated by the inhabitants of Tegea, and at the end of the fifth or beginning of the fourth centuries they passed a law stipulating that sacred animals on their way to a different place, perhaps in connection with transhumance, had a right to graze and rest for a day and night on the

---

308 *SIG*³ 963.
309 *LSCG* no. 105 = *IG* XII 5 2, dating from the fourth century.
310 Against the general opinion that the land in question was sacred see, not cogently, Jost 1985, 382.

sacred ground.³¹¹ If the period turned out to be protracted, it cost 1 drachma per day for a cow or a pig and 1 obol per day for a goat or sheep. The *hierothytes* (the man who arranged offerings to Alea), was allowed to have cattle suitable for offering graze on the area; perhaps this meant cattle belonging to the goddess, but this is not clear from the inscription. Her priest, personally, was allowed to graze 25 sheep, a couple of draught-animals, and a goat on the sacred area. They were his private animals, and if he allowed more animals than those to graze there, he was fined for each individual cow, whereas smaller animals were perhaps simply confiscated.³¹² Animals owned by private persons were otherwise allowed to graze on the area only when they had an errand for the sanctuary, either because they were to be sacrificed or because they pulled the cart for a worshipper. Finally, private persons in transit could be allowed to have their own team, but no other animals, graze in connection with a single overnight stay.

APOLLO'S ANIMALS IN DELPHI

We have seen how the large area below Delphi, once the land of the Kirrhaians, was to lie fallow and uninhabited. The general assumption is that the area could be used for grazing.³¹³ There is no indication to this effect in our written sources from the Classical period. In an important inscription from the middle of the fourth century, inspection of the area is ordered with a view to putting a stop to the illegal agricultural exploitation of the land by private people, and it is emphasized that the area is not to be fertilized nor inhabited. This is shown, for example, by virtue of the fact that no mill or mortar was allowed to be placed there.³¹⁴ This does not exclude the possibility that animals may have been able to pasture there.³¹⁵

No accounts of ordinary income from Delphi have been preserved. We do know that in 275 Apollo in Delphi received 50 oxen from a Lakedaemonian village as a present. The oxen were purchased in Lokris, probably to save transport. At that time, at least, Apollo probably had sacred cattle to which the 50 new ones were then to be added.³¹⁶ It is, however, only in an inscription from 178 that we have certain evidence to show that Apollo owned cows as well as horses. At that time, they were to graze within a more closely defined part of the sacred area where private persons were not permitted with their flocks (*thremmata*).³¹⁷

311 *IG* V 2.3; *LSCG* no. 67. For recent discussions see Guarducci 1952, Georgoudi 1974, Jost 1985, 382–5. For transhumance in general see pp. 99–101.
312 They were subjected to *inphorbismos*, the significance of which is not quite clear, cf. Guarducci 1952, 54–9. Sacred animals alone were excepted from this *inphorbismos*.
313 Bourguet 1905, 26; Kahrstedt 1953, 749.
314 *GD* 2.2501, 4.15–26, cf. Aischines 3.107–13.
315 But cf., concerning Chios, *SIG*³ 986 = *LSCG* 116.
316 *SIG*³ 407 with commentary.
317 *SIG*³ 636.

One might think that sacred animals were meant as a safeguard to ensure a supply of sacrificial animals there and then. This explanation may apply to the holy cows, but not to the horses. In the section on cattle-breeding we saw that the individual farmer, as a rule, *purchased* his horse if indeed he had one at all.[318] Perhaps Delphi was, or in the Hellenistic period became, simply one of the places where breeding of horses (and undoubtedly also cows) took place. If by sale the animals became privately owned, they lost their status of being sacred.

Apparently, by 117/16 the carelessness with the possessions of Apollo of Delphi had become too conspicuous. From the Romans, but on Greek initiative, the Amphiktyonians were told to take an inventory of Apollo's herds of animals to determine how much was owed to the god for the animals and by whom. But the Amphiktyonians gave up. All those questioned gave the answer that they did not know how many animals they had taken over, how many they sent on, or what had been the income from them. The inscription states that nothing was written in the public accounts about herds of animals.[319] Nevertheless, it appears that Apollo ought to have an income from his animals.

We cannot expect to find tenants of sacred land in Delphi except outside the Kirrhaian area.[320] In fact every mention of sacred animals at Delphi points towards animal-breeding without any connection with agriculture. The situation is quite different when one turns to Apollo's land on Delos.

APOLLO'S ANIMALS ON DELOS

On Delos the sacred land was to be cultivated intensively. Large parts of the accounts showing Apollo's ordinary income and expenses have been preserved, but one searches almost always in vain for mention of the sacred animals in the accounts.

We know that on Delos Apollo had his own animals, cows among others,[321] and that these animals grazed on Apollo's land. Yet in the accounts, tenants are listed only as tenants of land. In lists of what came with a piece of land when taken over by the tenant, buildings, vine and other fruit trees are mentioned,[322] but not the number of sacred animals, nor indeed anything to suggest that animals may have been included. Nor are animals mentioned as a separate category in connection with the official accounts.

In the fourth century there are scattered references to income for Apollo from the sale of wool from the sacred sheep.[323] Apart from that, incontes-

---

318 See pp. 87–9.
319 *SIG*³ 826 G.
320 For leasing of confiscated land see *SIG*³ 178, cf. 175 and Pomtow 1906.
321 For the opinion that Apollo owned only cows see Kent 1948, 293. Against this view Cavagnola 1973, 517 with note 28.
322 Olive was grown only on land acquired later, at Mykonos, *ID* 366 B.8–23, Kent 1948, 288.
323 *IG* II/III² 1639.15–16, 17; 1638.66; 1640.28.

table evidence to show that sacred animals were a permanent phenomenon may be derived from a revision of the rules guiding the leasing out of a god's land from about the year 300, the so-called *hiera syggraphe*.[324] Here there is a distinction between the animal-breeders and those who did not breed animals. Whereas the latter paid rent only once a year, in Metageitnion, corresponding to August–September time, the animal-breeders had to pay their rent in two rates. In Artemision, the Mounychion of the Athenians, our April–May, the first rate fell due. In the inscription, which is badly damaged, we read that at the time the animal-breeders had to pay a certain amount per animal. Apparently the animals referred to here were not sacred, but belonged to the tenant himself or to some other mortal.

Artemision is called Pokios (the 'Wool-month') in Lokris; that was the month when you sheared or plucked your sheep, and the month when lambs were old enough to be sold. It was also the time when the animals had to be transported from the island if there was not sufficient grass for them in the summer.[325] This must have been an additional motive for selling some of the animals.

Naturally, the animal-breeders derived a great part of their income at that time (see p. 91). The remainder of the rent was paid in Lenaion of the following year, the Gamelion of the Athenians, our January–February; this was the month when the herd of sheep and goats from the previous year were fully grown, and it was a suitable time to reduce the number of males among them. The females, on the other hand, had their lambs and kids to tend to.[326]

The rent to be paid by the animal-breeders was also fixed at a certain amount over a ten-year period with no regard to however many lambs, etc., might be born. The distribution of the rent into two rates is therefore usually looked upon as an assistance or insurance so that it fell due when the income was available. Besides, the temple may have had an interest in buying sacrificial animals at those times.[327]

The tenant's animals grazed together with those of Apollo. In the month before Artemision[328] the *hieropoioi*, according to the *hiera syggraphe* were to inspect 'the branded cows',[329] and perhaps the inspection was repeated in Metageitnion. This was probably also the occasion when animals of the new brood were branded.[330] The *hieropoioi* had to swear that the sacred animals which the tenant might now raise would have no influence on, or be

---

324 *ID* 503, dated to about 300 (Dürrbach 1919, 177–8) or to 290 (Kent 1948, 284–5).
325 There is no written evidence for this with respect to Delos, but cf. in general Robert 1949b.
326 Cf., on the sacrificial calendar of Erchia (see p. 176), Jameson 1988, 102–3.
327 The important festival called Apollonia was celebrated in the month Hieros, which followed Artemision. Apart from that, very little is known about the calendar of festivals at Delos, cf. Bruneau 1970, 86; 507–9.
328 Galaxion, corresponding to our March–April.
329 *ID* 503.21–2.
330 Cf. *IG* XI 2 287 A.44 and 58.

included in, the payment of the rent. This must mean that no fee was to be paid for them during the following month, but only for the animals in private ownership.[331]

As a rule branded animals were not for sale. If the particular tenant nevertheless wanted to sell one of Apollo's animals, he had to report it to the *hieropoioi*, and provide a guarantor for the price of the animal. In other words, the tenant was called upon to replace the animal sold with a new one that was then to be branded when the time came.[332]

The number of Apollo's animals

Offhand, it is likely that only limited herds of animals belonged to Apollo. The tenant took care of the branded sacred animals along with his own. In return, he probably had a right to the milk of the animals and perhaps to part of the wool as long as he took care of them. When a sacrificial animal was needed, the *hieropoioi* could have the animals fetched, and as there was no question of purchase or sale, it did not appear in their accounts. In the accounts of the *hieropoioi* for the year 250, which are the best preserved of all, only the hide of the ox and the goat sacrificed at the Poseidon Festival in the month of Posideon are listed as an income.[333] Under expenses only a corresponding expense connected with the purchase of animals remains unlisted; the reason for this may be that Apollo himself delivered the sacrificial animals that year.[334] But there are several other possibilities. Perhaps the sacrificial animals were financed by Delian citizens by a liturgy or from the amount of 600 drachmas that, during the period of independence, the Delian state seems to have yielded as an annual contribution to the Poseidon Festival.[335]

Each month an offering of purification of the sanctuary of Apollo was performed, and for this purpose each time a pig was purchased (as it appears from the accounts). In other words, Apollo himself did not own pigs, and this is understandable if the only advantage on the part of the tenant in looking after Apollo's animals was the benefit that he could derive from them while they were alive. In that case, he had no interest in tending pigs for Apollo; this he would have had if he had a right to the young of the sacred animals. The question is what benefit the tenant could derive from the cows. One could think, like Kent, of their milk and of their capacity as

331 *ID* 503.23–4. We give the usual interpretation of these difficult lines.
332 *ID* 503.25–7.
333 *IG* XI 2 287 A.24, cf. for the Apollonia Bruneau 1970, 65.
334 In one year, 179, the accounts include the purchase of a cow for the Posideia, *ID* 442 A. 219–22.
335 Bruneau 1970, 260–1. In the accounts of the Athenian Amphiktyonians covering the period 376/7–374/3 a purchase of 109 cows bought outside Delos is listed. This seems to have been for the Delia, which was not held during the period of independence, *SIG*³ 153, cf. Bruneau 1970, 81.

draught-animals. But cow's milk was not highly valued by the Greeks (see p. 89). Probably the tenants used the sacred cows as draught-animals, but hard worked draught-animals were not considered suited for sacrifice.[336]

Mixed agriculture on Delos?

There were only two ways in which the farmers on Delos and Rheneia, poor in plains as these islands are, could gain if not wealth then at least a secure livelihood: these were viticulture and animal-breeding, that is to say breeding of sheep. This is what is claimed by Bruno Cavagnola in his article from 1973. He presents a slightly misleading approach, which dims the much more varied picture of Delos and Rheneia. Its dimness is probably illustrated by the fact that on Delos rent is traditionally known as *enerosia*,[337] in other words, taxes on grain, and by the fact that according to the law the first thing to be done if a tenant is unable to pay his rent is to sell his harvest (*karpoi*). If that does not suffice, then his cows, goats and sheep, and his slaves, would have to be seized in execution. Further debt will be covered by selling what the tenant and guarantor might have.[338]

As animals do not appear in the accounts, we shall attempt to make deductions about their presence or absence by other means. The basic evidence is the occurrence of stables or shelters on several of Apollo's properties as mentioned in the accounts. *Boustasis* and *probaton* are terms used to denote stables designed for cows or oxen and smaller animals, respectively. In the accounts, cowsheds appear in thirteen of Apollo's properties; Charoneia had two cowsheds so that altogether fourteen cowsheds are attested by inscriptions, and twelve 'stables' for smaller animals, two of which belong to Charoneia.[339]

One might suppose that these stables were meant for animals which were imported and stayed on the island only for a brief period before they were to be slaughtered, primarily in connection with offerings. This does not agree with the *hiera syggraphe* from about the year 300, which we have mentioned above; there two types of tenants were involved. The evidence could still be dismissed if it were not for the decrees – admittedly, not fully preserved – that applied to the animal-breeders: the terms of payment and their placement in the year, the anticipated inspection once or, rather, twice a year.

The question is, therefore, to what extent mixed agriculture was involved; in other words, whether the animal-breeders also tilled their fields, in such a way that animal-breeding and agriculture were integrated. It is listed in the

---

336 Cavagnola 1973, 515, against Kent 1948, 293; cf. also Jameson 1988, 87 with note 3.
337 e.g. *IG* XI 2 162 (278).
338 *ID* 503. 33–6.
339 Kent 1948, 299–300.

leasing accounts how many vines, fig trees and other fruit trees were to be found on the area about to change tenants. It can be established that vine predominated, and that there were no olive trees on Delos itself although there were some on Mykonos, of which Apollo at one time received the western corner.

It is more difficult to determine how much grain and how much livestock were to be found on one property. One leasing area, the Hippodromos, had no fruit trees. Nor was there any *achyron* ('barn'), and as there was a cowshed as well as a paddock for sheep, it has been assumed that it was used for animal-breeding predominantly. Nevertheless, it appears that on one occasion at the end of the fourth century the *hieropoioi* of Apollo recovered part of the arrears of the Hippodromos tenant by 'selling the barley'. In addition two cows or oxen belonging to the tenant were sold. Out of the remaining amount the guarantor paid half whereas the tenant was listed as debtor of one and a half times the amount still outstanding. The lease was taken over by a new man, the son of the former tenant.[340]

Because of its name, the Hippodromos has been looked upon as something special – perhaps identical with the area that in the fifth century Nikias presented to Apollo, with the clause that every fourth year festival games should be held, and that, in connection with them, among other things, Nikias should be commemorated (see p. 183). At the same time, however, this example may also serve as a warning against deducing too much from the inventory of the accounts whenever we are not dealing with well-rooted plants like vine and fig.

The example of the Hippodromos may support a theory that there was a tendency towards more animal-breeding and less agriculture. The account of the debtor is earlier than the inventory first preserved. It may also render possible an interpretation that the cowsheds were filled with Apollo's own cows whereas the tenants had only a few. Finally, it may be viewed as a support for our explanation why animal-breeders were to pay twice a year. Individually, at odd times of the year, they could make their ability to pay deteriorate by selling some of their animals.

An argument against the assumption that Apollo's animals exceeded the others in number is the fact that the sacred estates which depended, according to Kent, primarily on animal-breeding and less on the production of wine managed better in the second half of the third century, which witnessed a decline in the rentals from the estates depending more on the production of wine.[341]

Osborne does not mention the sacred cows on Delos nor the possibility of the existence of other types of sacred animals on the island. In his interpretation of the *hiera syggraphe* he operates with a distinction between sheep-

---

340 *IG* XI 2 142.5–12 concerning this and another case of arrears.
341 Cf. Kent 1948, 309–10.

farmers and the others, not between animal-breeders and the rest. Consequently, he must assume that the sacred animals – whether they are the sacred cows mentioned in the inscription or possibly other types – were not of any real importance, and we have no means of telling to what extent he is right.

It seems to have usually happened that the tenant moved at the end of the ten-year period. In Osborne's opinion this supports the assumption that although mixed agriculture did exist on the island, it was sheep-breeding that was of economic importance for the tenants: sheep and slaves could easily be moved to a new piece of land, and the wealthy tenant himself resided inside the city.[342] At any rate, it is not in disagreement with Pečírka's assumption[343] that those who lived in the country were those who had no business in the city – in this case primarily the slaves who managed the farming, slaves who could be sold if the tenant could not pay his dues.[344]

At this point we may conclude that very little material on sacred animals is available, especially from the Classical period. This *may* correspond to the extent of the phenomenon, but it may also be a reflection on the method of accounting. Systematic rearing and perhaps breeding of animals owned by gods is attested with certainty concerning the major sanctuaries consecrated to Apollo. A prerequisite for a god to own animals on a more permanent basis was undoubtedly that the god himself owned land to keep them. The material is too scant to show whether it was in particular costly animals like cows and horses that were owned by the gods. Delos and Rheneia were small islands and cannot have supported an unlimited number of cows. Still it may be concluded that here, irrespective of Apollo's share in the number of animals, we are very close to a system that may look like mixed agriculture. The reason for this must be, in particular, the constant demand for sacrificial animals entertained by the sanctuaries of the islands, first and foremost the temple of Apollo.

---

342 Osborne 1989, 301–2. Against this interpretation of the epigraphic evidence see Brunet 1990.
343 Pečírka 1973, 118–19.
344 Kent 1948, 280 with note 129.

EPILOGUE

This study has no conclusion. The presentation of various aspects of ancient Greek agriculture does not yield a simple and clear overall picture. You might say that the most characteristic feature is that of variety. This is due to the variety of the landscape and the position of the arable areas, often scattered in small pockets, and to the well-known political and social differences between the different city-states.

From a technical point of view agriculture seems to have been rather primitive. The different tools are simple and often made locally by the farmer himself. Only the technologies for making olive-oil and wine seem to show some progress. This does not mean that one cannot produce a considerable surplus provided one has the labour force necessary, be it chattle-slaves or helots or other types of semi-free populations. In this respect also the Greek city-states show a great many variations, as far as they are known to us.

The main problem is to establish what is normal, and we have tried to show how difficult it is to arrive at any certainty on this point. One might argue that this is due partly to the rather restrictive use of the sources. The book is, in fact, primitivistic and deliberately in opposition to certain modern views that we would not hesitate to call modernistic, using another interpretation of the sources, especially of Theophrastus,[345] and introducing later sources from the Roman period, especially concerning Italian agriculture. We have confined ourselves to the use of contemporary evidence, and for that reason one might, therefore, be justified in accusing us of *Quellenpositivismus* (quod non est in actis non est in re). Here we should like to stress that this book is an introduction. The aim is to establish what can be seen from the contemporary sources, be they archaeological or literary, and confront them with the given facts of climate, precipitation and – to some extent – modern non-industrialized experience from agriculture in contemporary Greece. This must be the basis and we hope that others will continue these efforts to understand the agricultural systems better and in a

---

345 So, cautiously, Garnsey in Wells, ed., 1992.

more varied way. We do not expect – nor do we hope – that this will be the last word.

On the other hand, we must consider that our rather pessimistic evaluation of agriculture as the most important source of wealth requires a more elaborate model for the economy of the ancient Greek city-state than that of the Finley school.[346] We know that most of the city-states, which numbered over a thousand, were more like tiny towns or hamlets, and the economy of such centres was, of course, somewhat rural. On the other hand, the economy was a curious mixture of gift-giving and a market economy. Taxes were, as we have seen, very restricted, but people had to pay something and the state was entirely dependent on rich people who would gain honour and prestige by lavish expenditures for the common weal. This was systematized in different ways in the different city-states among which the best known is Athens, with her elaborate system of *leitourgiai* and complicated sets of rules for those who were expected to undertake the burdens. We shall not go into details but only stress that the rich had to pay many of the necessities that, to a modern mind, are public expenditures, and they had to pay in cash. We do not believe that the rich derived their wealth from the sale of their own agricultural products to any great degree. Prerequisites for this would have been both concentration of the land in the hands of the rich and more favourable conditions for cultivation, both as regards labour force and technology.

Trade in other products, including foreign agricultural products for import, might be one important source of income. On the other hand we have only a few traces of large-scale trade. The reason might well be that trade was organized differently. The cargo of the single ship often consisted of different loads owned by different people. Just as splitting up land in small plots was a division of risks, so the divided ownership of a cargo minimized the risk of losing everything, if each owner had parts of many cargoes. Even the most honoured of the semi-public expenditures, the Athenian trierarchy, was divided into smaller parts from the end of the fifth century until the fall of the Athenian state as a seapower at the end of the fourth.[347] Dividing the risk was a common practice.

In his description of Levantine trade in the seventeenth century, Steensgaard (1973) has used the term 'peddling trade', and although we do not venture to make any more serious comparison between the city-states of Ancient Greece and early modern Levantine trade, it is a possible solution of the problem of the rich man's economy that he had divided his investments in many enterprises, drawing his ready money from many different sources. We should not wonder why this is not mentioned more often in the sources, the peddlers normally being very reticent about their trade and often acting as agents for others. The rather obscure situation described in the famous

---

346 See especially Finley 1973a.
347 Gabrielsen 1991 discusses the trierarchic institution from many points of view.

letter from Berezan is revealing. A tradesman in difficulty calls in a more powerful man to get protection for his person and his cargo. Perhaps the stronger party is the real owner.[348] We do not possess many documents of this type, but its existence in the territory of a Greek colony in a corn-exporting area before 500 BC is important. Many centuries later, Dio Chrysostomus describes the trade in the same area. Discussing Homer in relation to other poets he introduces a simile:

> Just as when a merchant (*tis ton emporon*) sails into your port who has never been there before you do not immediately scorn him but, on the contrary, having first tasted his wine and sampled any other merchandise in his cargo, you buy it if it suits your taste, otherwise you pass it by.
>
> (Dio Chrysostomus 36.11)

The situation is so banal that it is scarcely worth mentioning, but we should like to stress the expression 'one of the merchants' (that is, the situation is a typical one) and combine this with Dio's general description a little earlier in the same speech, when he says that at this time Olbia is only a port of trade for the Scythians in the hinterland.

Of course, it was possible to grow rich from agriculture, but prices were extremely variable, and the risk of failure of crops considerable; you would have to control a considerable area of arable land and a large labour force.[349] Other sources of income could easily be combined with agriculture, hiring out oxen for transport, selling products from the farmstead abroad – which to an Athenian could well be olive-oil for Aigina or Megara – investing in shipping and corn-trade, mining or quarrying, peddling trade via agents, and so on.

The ever-increasing number of containers, especially amphoras with or without stamps, is an important indication with regard to trade involving agricultural products, especially olive-oil and wine. Most of the evidence is to be dated to the Hellenistic-Roman period and we cannot use this material directly for the earlier periods. A short introduction to the amphoras and the ancient wine trade is given by Virginia R. Grace (1979). The substantial export of olive-oil from Athens and the so-called SOS amphoras from the seventh and sixth centuries testify to a trade of some extent but we cannot evaluate the role of this trade for the economy.[350] It has been argued that the export of fine ceramics in bulk was very limited, but when you consider that vases were only a small part of a cargo, they testify to a much more extensive

---

348 Austin/Vidal-Naquet 1977, 220–3, following the text edn. by Bravo (1974) with ample commentaries.
349 The role of agriculture as a source of wealth was, of course, discussed by many participants at the symposium on Agriculture in Ancient Greece, see Wells, ed., 1992.
350 Johnston/Jones 1978, Baccarin 1990 with ample references. We cannot follow his rather bold statements. For further discussion of Greek amphoras see Empereur/Garlan 1986, and especially for Thasian amphoras, Salviat 1986.

trade that touched upon nearly every inhabitated place in the Mediterranean world. Therefore, there must have been a lot of other goods constituting the bulk of trade. Mixed cargoes are also attested from the finds of shipwrecks,[351] and we can guess at the extent when we examine the rules for maritime loans.[352]

We should prefer to characterize the ancient Greek economy as a *mixed economy*. It never arrived at a higher level, and the reason might well be that the *leitourgiai* took much of the rich man's surplus which in a capitalistic mode of production would have been invested. The small city-states with an upper class who used conspicuous semi-public consumption to acquire status and a lower class that expected the rich to do so have limits of economic growth.

351 Parker 1984.
352 Isager/Hansen 1975.

APPENDIX: THE SACRED OLIVES

While in Eleusis, Demeter gave grain to the whole world, the olive was Athene's gift especially to the Athenians. The olive that, according to the myth, she let spring forth on the Acropolis is probably to be viewed as the first olive of the world.[353] The Athenians regarded the olive that was in fact growing on the Acropolis as the sacred tree mentioned in the myth, and twelve sacred olives that grew on the ground of the Academy in the Classical period were looked upon as products of cuttings from that tree.

Bundgaard (1976) discusses the relation between the Erechtheion[354] and the sacred olive tree, placing the tree inside the existing temple of the fifth century. We shall not enter into this discussion but only point out that this tree could, in fact, well be the first olive in Attica. According to Herodotus (8.55) the tree caught fire during the Persian destruction of the Acropolis in 480, but 'when the Athenian exiles at the king's order went to the sanctuary they saw that a shoot of about a cubit's length had sprung from the bole'.

Olive trees are difficult to destroy and even if the story is a little exaggerated it seems likely that the sacred olive was not a pruned one but a genuine domesticated olive. It could well have been imported as a young tree and planted on the sacred area of the Mycenaean city perhaps a thousand years before the Persian wars. As Sophocles (*Oedipus Coloneus*, 695 ff.) describes the tree as enormous the original trunk could hardly have been destroyed. It is worth noting that he uses the word *paidotrophos* – that is, the tree has many children, obviously root shoots and cuttings transplanted to other places. In that case we have a mythical explanation of the arrival and the spreading of the olive tree in Attica which could very well reflect an historical event.

The sacred olives in Attica were called *moriai*, and this term was applied not only to the trees in the Academy, but also to sacred olives growing about in Attica. The name has a connotation denoting something like 'part of', and

---

353 Herodotus 5.82 indicates that it was part of the myth.
354 The name 'Erechtheion' here refers to the building that has enjoyed that name among modern scholars until recently when doubt was raised by Kristian Jeppesen, cf. especially Jeppesen 1987.

this may be a reference to the circumstance that the trees were or were at least regarded as cuttings from the holy tree on the Acropolis. A different, but less attractive, theory will have it that the trees were called *moriai* because they had been chosen to constitute Athene's share of the olives in Attica.[355]

In Attica there were regulations concerning the removal of any olive tree. The law as cited in Demosthenes' speech *Against Makartatos* says:

> If anyone shall dig up an olive tree at Athens, except it be for a sanctuary of the Athenian state or of one of its demes, or for his own use to the number of two olive trees each year, or except it be needful to use it for the service of one who is dead, he shall be fined 100 drachmae, to be paid into the public treasury, for each olive tree.
> (Demosthenes 43.71)[356]

Presumably this is to indicate that in general terms no one should detract from the value of the land in consideration of his descendants. While a fine was imposed on people who broke this law, digging up one of Athene's sacred olives was originally punished by death. From ancient times landowners on whose land such trees were grown had a duty to deliver each year one and a half *kotyle* – that is, three quarters of a litre of oil from each of the sacred trees – to the archon. At the time of Aristotle it was no longer checked from which tree the oil was derived, and the law that called for the death penalty for destruction of a sacred olive was no longer in use.[357] Judging by Lysias' speech about the olive stump it seems that then, shortly after the year 397, the punishment was exile and confiscation of property.[358]

Athene's oil was to be used as a prize of victory at the Panathenaia, and probably only at the Great festival, which was celebrated every fourth year. The fact that jars inscribed with names of archons have been found shows that the archon's responsibility for the collection of the oil was taken seriously. Probably the name of the archon indicates the year when the amphora was sealed with freshly pressed oil. When presenting his accounts, the archon had to display a definite number of amphorae bearing his name, and he could not take his seat in the Council of Areopagus until the amphorae had been delivered on the Acropolis. For the winner of the prize, the sealing by the archon had the advantage that he could tell how old the oil was. Theoretically, it could be derived from one of the preceding four archon-years and not from the year when the prize was to be awarded, as the festival took place at midsummer in the first month of the year,

---

355 This is the opinion of Latte 1933, 302.
356 Translated by A.T. Murray, Loeb edn.
357 Aristotle, *Ath. Pol.* 60. Other laws no longer used were that on ostracism and that requiring membership of one of the upper Solonian classes for certain magistrates.
358 Lysias 7.3, 32, 41.

APPENDIX

Hekatombaion. Just how old the oil was before the Greeks considered it to be too old, we have no way of telling.[359]

But there is no reason to believe that the oil grew rancid in the home of the victorious athlete who had won perhaps fifty Panathenaïc amphorae filled with sacred oil. While the export of common olive-oil was forbidden the sealed prize amphorae could probably be exported and their contents sold. Some of them ended up in sancturies but there is no way of telling as yet if they were full or empty when they were given to the god.[360]

---

359 cf. Gardiner 1912 and Rhodes 1981 on Aristotle, *Ath. Pol.* 60.
360 Valavanis 1986, with ample references.

BIBLIOGRAPHY

Adamesteanu, D. (1973), 'Le suddivisioni di terra nel Metapontino', in Finley, M.I., ed., *Problèmes de la terre en Grèce ancienne*, Paris, 49–61.
—— (1974), *La Basilicata antica. Storia e monumenti*, Di Mauro.
Amouretti, M.C. (1985), 'La transformation des céreales dans les villes, un indicateur méconnu de la personnalité urbaine. L'éxemple d'Athènes à l'époque classique', in Leveau, Ph., ed., *L'Origine des richesses dépensées dans la ville aitique*, Aix-en-Provence, 133–46.
—— (1986), *Le Pain et l'huile dans la Grèce antique* (Annales Littéraires de l'Université de Besancon 328), Paris.
Amouretti, M.C./Comet, G. (1985a), *Le Livre de l'olivier*, Aix-en-Provence.
—— (1985b), 'Iconographie et histoire des techniques', in *Histoire des techniques et sources documentaires, méthodes d'approche et expérimentation en région méditerranéenne* (CNRS-GIS, Maison de la Méditerranée. Cahier no. 7), 207–17.
Amouretti, M.C./Comet, G./Ney, Cl./Paillet, J.L. (1984), 'A propos du pressoir à huile: de l'archéologie industrielle à l'histoire', *Mélanges de l'Ecole Française de Rome*, 'Antiquité' series, 379–421.
Amyx, D.A. (1958), 'The Attic Stelae. Part III', *Hesperia* 27, 163–310.
Andreades, A. (1965), *Geschichte der griechischen Staatswirtschaft*, Hildesheim. (First edn, 1931.)
Andreyev, V.N. (1974), 'Some aspects of agrarian conditions in Attica in the fifth to third centuries BC', *Eirene* 12, 5–46.
—— (1983), 'Zur Kontinuität der Vermögenselite Athens vom 5. bis 3. Jahrhundert v.u.Z.', *JWG* 137–58.
Argoud, G. (1981), 'L'alimentation en eau des villes grecques', in *L'Homme et l'eau en Méditerranée et au Proche–Orient, I, Séminaire de recherche 1979–1980*, Lyons, 69–82.
'Army Forces Service Manual, Greece' (1943), unpublished ms.
Aschenbrenner, S. (1972), 'A contemporary community', in McDonald, W.A./Rapp, G.R., eds, *The Minnesota Messeria Expedition. Reconstructing a Bronze Age Environment*, St Paul/Oxford/Toronto, 47–63.
Asheri, D. (1961), 'Sulla legge di Epitadeo', *Athenaeum* 39, 45–68.
—— (1963), 'Laws of inheritance, distribution of land and political constitutions in ancient Greece', *Historia* 12, 1–21.
—— (1965), 'Distribuzioni di terre e legislazione agraria nella Locride occidentale', *JJP* 15, 313–28.
—— (1966), *Distribuzioni di terre nell'antica Grecia* (MAT IV.10), Turin.
—— (1971), 'Supplementi coloniari e condizione giuridica della terra nel mondo greco', *Rivista di Storia Antica* I, 77–91.

BIBLIOGRAPHY

Audring, G. (1973), 'Uber den Gutsverwalter (*epitropos*) in der attischen Landwirtschaft des 5. und des 4. Jh. v.u.Z.', *Klio* 55, 109–16.
—— (1985), 'Zur sozialen Stellung der Hirten in archaischer Zeit. Thesen', in Kreissig, H./Kühnert, F., eds, *Antike Abhängigkeitsformen in der griechischen Gebieten ohne Polisstruktu und den römischen Provinzen* (Schriften zur Geschichte und Kultu der Antike 25), Berlin, 12–19.
Austin, M.M./Vidal-Naquet, P. (1977), *Economic and Social History of Ancient Greece: An Introduction*, trans. and rev. by M.M. Austin, Berkeley/Los Angeles.
Baccarin, A. (1990), 'Olivicoltura in Attica fra VII e V sec. a.C. Trasformazione e crisi', *Dialoghi di Archeologia* (3rd series) 8.1, 29–33.
Bakhuizen, S.C. (1975), 'Social ecology of the ancient Greek world', *Antiquité Classique* 44, 211–18.
Barron, J.B. (1983), 'The fifth-century *horoi* of Aigina', *JHS* 103, 1–12.
Beazley, J.D. (1956), *Attic Black-Figure Vase-Painters*, Oxford.
Behrend, D. (1970), *Attische Pachturkunden: Ein Beitrag zur Beschreibung der misthosis nach den griechischen Inschriften* (Vestigia 12), Munich.
Bell, M./Limbrey, S. (1982), eds, *Archaeological Aspects of Woodland Ecology*, Oxford.
Bergier, J.-F. (1989), ed., *Montagnes, fleuves, forêts dans l'histoire. Barrières ou lignes de convergence*, St Katharinen.
Berthiaume, G. (1982), *Les Rôles du mágeiros. Etude sur la boucherie, la cuisine et le sacrifice dans la Grèce ancienne* (*Mnemosyne* Suppl. 70), Leiden.
Bettalli, M. (1982), 'Note sulla produzione tessile ad Athene in età classica', *Opus* 1, 261–78.
Bevan, E. (1986), *Representations of Animals in Sanctuaries of Artemis and Other Olympian Deities* (BAR 315), Oxford.
Bickerman, E.J. (1980), *Chronology of the Ancient World*, London. (First edn, 1968.)
Billiard, R. (1913), *La Vigne dans l'antiquité*, Lyons.
Bintliff, J.L./Snodgrass, A.M. (1985), 'The Cambridge/Bradford Boeotian Expedition: the first four years', *Jl Field Arch.* 12, 123–61.
Blanck, H. (1980), 'Essen und Trinken bei Griechen und Römern', *Antike Welt* 11(1), 17–34.
Boardman, J. (1956), 'Delphinion in Chios', *BSA* 51, 41–54.
—— (1958–9), 'Excavations at Pindakas in Chios', *BSA* 53/54, 295–309.
—— (1967), *Greek Emporio* (*BSA* Suppl. 6), London.
Boardman, J./Vaphopoulou-Richardson, C.E. (1986), eds, *Chios: A Conference at the Homereion in Chios 1984*, Oxford.
Bockisch, G. (1985), 'Die Helotisierung der Messenier. Ein Interpretationsversuch zu Pausanias 4.14,4 f', in Kreissig, H./Kühnert, F., eds, *Antike Abhängigkeitsformen in der griechischen Gebieten ohne Polisstruktu und den römischen Provinzen* (Schriften zur Geschichte und Kultu der Antike 25), Berlin, 29–48.
Bodson, L. (1983a), 'Aperçu de l'élevage bovin dans l'antiquité,' *Ethnozootechnie* 32, 38–50.
—— (1983b), 'Attitudes toward animals in Greco-Roman Antiquity', *International Journal for the Study of Animal Problems* 4, 312–20.
Bolkenstein, H. (1958), *Economic Life in Greece's Golden Age*, Leiden. (First edn, 1923.)
Bolla, S. von (1940), *Untersuchungen zur Tiermiete und Viehpacht im Altertum* (Münchener Beiträge zur Papyrusforschung 30), Munich.
Bosanquet, R.G. (1902), 'Excavations at Praesos I', *BSA* 8, 231–70.
Boserup, E. (1965), *Conditions of Agricultural Growth*, London.

BIBLIOGRAPHY

Bouras, C. (1984), *Chios: Greek Traditional Architecture*, Athens.
Bourguet, E. (1905), *L'Administration financière du sanctuaire pythique au IVe siècle avant J.C.*, Paris.
Bousquet, J. (1965), 'Convention entre Myania et Hypnia', *BCH* 89, 665–81.
Boyd, Th.D./Jameson, M.H. (1981), 'Urban and rural land division in Ancient Greece', *Hesperia* 50, 327–42.
Boyd, Th.D./Rudolph, W.W. (1978), 'Excavations at Porto Cheli and vicinity', *Hesperia* 47, 333–55.
Bradford, J.M.A. (1956), 'Fieldwork on aerial discoveries in Attica and Rhodes', *AntJ* 36, 57–69; 172–80.
—— (1957), *Ancient Landscapes. Studies in Field Archaeology*, London.
Braudel, F. (1949), *La Méditerranée et le monde méditerranéen à l'époque de Philippe II*, 2 vols, Paris. (Transl. into English by Reynolds, S. (1973, 1975), London.)
Bravo, B. (1974), 'Une lettre sur plomb de Berezan. Colonisation et modes de contact dans le Pont', *Dialogues d'Histoire Ancienne* 1, 110–87.
Brendel, O. (1934), *Die Schafzucht im alten Griechenland*, diss. Giessen.
Bruhns, H. (1985), 'De Werner Sombart à Max Weber et Moses I. Finley: La typologie de la ville antique et la question de la ville de consommation', in Leveau, Ph., ed., *L'Origine des richesses dépensées dans la ville antique*, Aix-en-Provence, 255–69.
Bruneau, Ph. (1970), *Recherches sur les cultes de Délos à l'époque hellénistique et à l'époque impériale*, Paris.
Bruneau, Ph./Fraisse, Ph. (1981), 'Un pressoire à vin à Délos', *BCH* 105,128–69.
Brunet, M. (1990), 'Contribution à l'histoire rurale de Délos aux époques classique et hellénistique', *BCH* 114, 669–82.
Brunšmid, J. (1898), *Die Inschriften und Münzen der griechischen Städte Dalmatiens. Abhandlungen der archäologisch–epigraphischen Seminares der Universität Wien* 13, Vienna.
Bryer, A. (1986), 'Byzantine agricultural implements. The evidence of Medieval illustrations of Hesiodos' Works and Days', *BSA* 81, 45–85.
Bugh, G.R. (1988), *The Horsemen of Athens*, Princeton.
Bundgaard, J.A. (1976), *Parthenon and the Mycenean City on the Heights*, Copenhagen.
Burford, A. (1960), 'Heavy transport in classical antiquity', *Economic History Review* (2nd series) 13, 1–18.
Burford Cooper, A. (1977/8), 'The family farm in Greece', *CJ* 73, 162–75.
Burkert, W. (1977), *Griechische Religion der archaischen und klassischen Epoche*, Stuttgart.
—— (1979), *Structure and History in Greek Mythology and Ritual*, Berkeley.
Carlier, P. (1977), 'La vie politique à Sparte sous le règne de Cléomène 1er. Essai d'interprétation', *Ktema* 2, 65–84.
Carlsen, J. (1990), 'Et spektrum af afhængighedsformer', *Den Jyske Historiker* 51/52, 123–36.
Carpenter, J./Boyd, T. (1977), 'Dragon houses: Euboia, Attika, Karia', *AJA* 81, 179–215.
Cartledge, P.A. (1979), *Sparta and Lakonia. A Regional History 1300–362 BC*, London/Boston/Henley.
—— (1981), 'Spartan wives: liberation or licence?', *Classical Quarterly* 31, 84–105.
Cartledge, P.A./Harvey, F.D. (1985), eds, *Crux: Essays Presented to G.E.M. de Ste Croix on his 75th Birthday* (*History of Political Thought* 6. 1/2), Exeter.
Cartledge, P.A./Spawforth, A.J.S. (1989), *Hellenistic and Roman Sparta. A Tale of Two Cities*, London/New York.

BIBLIOGRAPHY

Cary, M. (1949), *The Geographic Background of Greek and Roman History*, Oxford.
Casabona, J. (1966), *Recherches sur le vocabulaire des sacrifices en grec des origines à la fin de l'époque classique*, Aix-en-Provence.
Casanova, G. (1981), 'I contratti d'affitto fra privati nelle epigrafi greche', in Bresciani, E./Geraci, G./Pernigotti, S./Susini, G., eds, *Scritti in onore di Orsolina Montevecchi*, Bologna, 89–97.
Cassola, F. (1965), 'Sul l'alienabilità del suolo nel mondo greco', *Labeo* 11, 206–19.
Cataldi, S./Moggi, M./Nenci, G./Panessa, G. (1981), *Studi sui rapporti interstatali nel mondo antico*, Pisa.
Catardi, C. (1973), 'Osservazioni circa il regime fondiario peloponnesiaco dalla fine del IV° sec. al II° sec. a.C.', *Istituto Lombardo (Rend. Lett.)* 107, 502–10.
Cataudella, M.R. (1976), 'Aspetti del diritto agrario greco: l'affrancazione', *Jura* 27, 88–101.
Cavagnola, B. (1972), 'I locatari delle proprietà fondiarie del dio Apollo a Delo', *Istituto Lombardo (Rend. Lett.)* 106, 51–115.
—— (1973), 'Aspetti economici dell'allevamento a Delo e Reneia in età ellenistica', *Istituto Lombardo (Rend. Lett.)* 107, 511–45.
Cavanaugh, M.B. (1980), *Eleusis and Athens: Documents in Finance, Religion and Politics in the Second Half of the Fifth Century BC*, Ann Arbor.
Chandor, A. Brisbane (1976), 'The Attic Festivals of Demeter and their Relation to the Agricultural Year', diss. University of Pennsylvania.
Cherry, J.F. (1988), 'Pastoralism and the role of animals in the pre- and proto-historic economies of the Aegean', in Whittaker, C.R., ed., *Pastoral Economies in Classical Antiquity* (Cambridge Philological Society Suppl. 14), Cambridge, 6–34.
Christiansen, J./Melander, T. (1988), eds, *Proceedings of the Third Symposium on Ancient Greek and Related Pottery*, Copenhagen.
Clinton, K. (1974), *The Sacred Officials of the Eleusinian Mysteries (Transactions of the American Philosophical Society* 64.3), Cleveland.
Cloché, P. (1931), *Les classes, les métiers, le trafic*, Paris.
Clutton-Brock, J./Grigson, C. (1984), 'Animals and archaeology', *BAR* 4, 227.
Coarelli, F. (1981), 'La Sicilia tra la fine della guerra Annibalica e Cicerone', in Giardina, A./Schiavone, A., eds, *L'Italia: Insediamenti e forme economiche*, Bari, vol. 1, 1–18.
Coldstream, J.N. (1977), *Geometric Greece*, London.
Condurachi, E. (1968), 'Problemi della *polis* e della *chora* nelle città greche del Ponto Sinistro,' in *La città e il suo territorio. Atti del Settimo Convegno di Studi sulla Magna Grecia, Taranto 8–12 ottobre 1967*, Naples, 143–63.
Craik, E.M. (1990), ed., *Owls to Athens. Essays on Classical Subjects Presented to Sir Kenneth Dover*, Oxford.
Culley, G.R. (1975), 'The restoration of sanctuaries in Attica. IG II^2 1035', *Hesperia* 44, 207–23.
—— (1977), 'The restoration of sanctuaries in Attika, II. The structure of IG II^2 1035 and the topography of Salamis', *Hesperia* 46, 282–98.
Dakaris, S. (1982), 'Von einer kleinen ländlichen Ansiedlung des 8.-4. Jhs. v. Chr. zu einer spätklassischen Stadt in Nordwest-Griechenland', in Papenfuss, D./Strocka, V.M., eds, *Palast und Hütte*, Mainz am Rhein, 357–74.
Damsgaard–Madsen, A./Christiansen, E./Hallager, E. (1988), eds, *Studies in Ancient History and Numismatics Presented to Rudi Thomsen*, Aarhus.
Daux, G. (1926), 'Nouvelles inscriptions de Thasos', *BCH* 50, 213–49.
—— (1963), 'La grande démarchie: un nouveau calendrier sacrificiel d'Attique (Erchia)', *BCH* 87, 603–34.
—— (1983), 'Le Calendrier de Thorikos au Musée J. Paul Getty', *L'Antiquité*

Classique, 150–74.
Davies, J.K. (1977), 'Athenian citizenship: the descent group and the alternatives', *Classical Journal* 73, 105–21.
—— (1981), *Wealth and the Power of Wealth in Classical Athens*, New York.
Davis, J.L./Cherry, J.F./Mantzourani, E. (1985), 'An archaeological survey of the Greek island of Keos', *National Geographic Society Research Reports* 21, 109–16.
Dawkins, R.M. (1929), *The Sanctuary of Artemis Orthia at Sparta, Excavated by the British School of Athens*, (*JHS* Suppl. 5), London.
de Ste Croix, G.E.M. (1966), 'The Estate of Phaenippus (Ps.-Dem. 42)', in *Ancient Society and Institutions: Studies Presented to Victor Ehrenberg*, Oxford, 109–14.
—— (1981), *The Class Struggle in the Ancient Greek World*, London.
Détienne, M. (1964), *Crise agraire et attitude religieuse chez Hésiode* (*Coll. Latomus* 68), Brussels.
—— (1973), 'L'olivier: un mythe politico-religieux', in Finley, M.I., ed., *Problèmes de la terre en Grèce ancienne*, Paris, 293–306 (= *Revue de l'Historie des Religions* 3 (1970), 5–23).
Détienne, M/Vernant, J.P. (1979), eds, *La Cuisine du sacrifice en pays grec*, Paris.
Deubner, L. (1932), *Attische Feste*, Berlin. (Second edn, 1966.)
Diels, H. (1965), *Antike Technik*, reprint of 3rd edn, 1924, Osnabrück.
Dimen, M./Friedl, E. (1976), eds, *Regional Variation in Modern Greece and Cyprus: Towards a Perspective on the Ethnography of Greece* (Annals New York Academy of Sciences 268), New York.
Di Stefano, G. (1980/1), 'Ricerche nella Provincia di Ragusa', *Kokalos* 26/27, 756–63.
Dohm, H. (1964), *Mageiros. Die Rolle des Kochs in der griechisch-römischen Komödie* (Zetemata 32), Munich.
Dörpfeldt, W. (1895), 'Die Ausgrabungen am Westabhange der Akropolis', *AM* 20, 161–206.
Dow, S. (1965), 'The Greater Demarkhia of Erkhia', *BCH* 89, 180–213.
—— (1968), 'Six Athenian sacrificial calendars', *BCH* 92, 170–86.
Dow, S./Healey, R.F. (1965), *A Sacred Calendar of Eleusis* (*Harvard Theological Studies* 21), Cambridge, Mass.
Drachmann, A.G. (1932), *Ancient Oil Mills and Presses* (Danske Vidensk. Selskab, Arch.-Kunsthist. Medd. I.1), Copenhagen.
—— (1938), 'Pflug', *RE* 21, 1461–72.
Du Boulay, J. (1974), *Portrait of a Greek Mountain Village*, Oxford.
Ducat, J. (1990), *Les Hilotes* (*BCH* Suppl. 20), Paris.
Dufková, M./Pečírka, J. (1970), 'Excavations of farms and farmhouses in the chora of Chersonesos in the Crimea', *Eirene* 8, 123–74.
Dunbabin, T.J. (1940), ed., *Perachora, The Sanctuaries of Hera Akraia and Limeria. Excavations of the British School of Archaeology of Athens, 1930–3*, Oxford, vol. 1.
—— (1957), *The Greeks and their Eastern Neighbours* (*JHS* Suppl. 8), London.
Dunst, G.(1977), 'Der Opferkalender des attischen Demos Thorikos', *ZPE* 25, 243–64.
Dürrbach, F. (1919), 'La *hiera syggraphe* de Délos', *REG* 32, 167–78.
Dusanic, S. (1978), 'Notes épigraphiques sur l'histoire arcadienne du IVe siècle', *BCH* 102, 333–58.
Ehrhardt, N. (1983), *Milet und seine Kolonien. Vergleichende Untersuchung der kultischen und politischen Einrichtungen*, Frankfurt am Main/Bern/New York.
Einarson, B. (1976), *Theophrastus, De Causis Plantarum I*, Loeb edn, New York/London.
Empereur, J.-Y./Garlan, Y. (1986), eds, *Recherches sur les amphores grecques* (*BCH* Suppl. 13), Paris.

Etienne, R. (1985a), 'Les femmes, la terre et l'argent à Ténos à l'époque hellenistique', in *La Femme dans le monde méditerranéen. I. Antiquité* (Travaux de la Maison de l'Orient 10), 61–70.

—— (1985b), 'Les sources de revenus des Déliens à l'époque hellenistique', in Leveau, Ph., ed., *L'Origine des richesses dépensées dans la ville antique*, Aix-en-Provence, 47–67.

Faure, P. (1978), *La Vie quotidienne des colons grecs de la mer Noire à l'Atlantique au siècle de Pythagore, VIe siècle avant J.-C.*, Paris.

Felsman, H. (1937), *Beiträge zur Wirtschaftsgeschichte von Delphi*, diss. Hamburg.

Feyel, M. (1936a), 'Etudes d'épigraphie béotienne I', *BCH* 60, 175–83.

—— (1936b), 'Etudes d'épigraphie béotienne II' *BCH* 60, 389–415.

—— (1937), 'Etudes d'épigraphie béotienne III', *BCH* 61, 217–35.

—— (1942), *Contribution à l'épigraphie béotienne*, Le Puy.

Figueira, T.J. (1984), 'Mess contributions and subsistence at Sparta', *TAPA* 114, 87–109.

Finley, M.I. (1968a), 'Sparta', in Vernant, J.P., ed., *Problèmes de la guerre en Grèce ancienne à la memoire d'André Aymard*, Paris, 143–60 (= Finley, M.I. (1975), *The Use and Abuse of History*, London/New York, 161–77).

—— (1968b), 'The alienability of land in ancient Greece: a point of view', *Eirene* 7, 25–32 (= Finley, M.I. (1975), 153–60; in French (1970): *Annales (ESC)* 25, 1271–7).

—— (1973a), *The Ancient Economy*, London.

—— (1973b), ed., *Problèmes de la terre en Grèce ancienne*, Paris.

—— (1973c), *Studies in Land and Credit in Ancient Athens 500–200 B.C.: The Horos-Inscriptions*, New Brunswick/New Jersey. (First edn, 1952.)

—— (1975), *The Use and Abuse of History*, London/New York.

Fitton Brown, A.D. (1984), 'The contribution of women to ancient Greek agriculture', *LCM* 9, 71–4.

Forbes, H.A. (1976a), 'The "thrice–ploughed" field: cultivation techniques in ancient and modern Greece', *Expedition* 19.1, 5–11.

—— (1976b), 'We have a little of everything', in Dimen, M./Friedl, E., eds, 236–50.

Forbes, H.A./Foxhall, L. (1978), 'The Queen of All Trees', *Expedition* 21.1, 37–47.

Forbes, M.H.C. (1976) 'Gathering in the Argolid: a subsistence subsystem in a Greek agricultural community', in Dimen, M./Friedl, E., eds, 251–64.

Forbes, R.J. (1964–66), *Studies in Ancient Technology*, 9 vols, Leiden.

Forrest, W.G. (1980), *A History of Sparta 950–192 B.C.*, London. (First edn, 1968.)

Fox, R.L. (1985), 'Aspects of inheritance in the Greek world', in Cartledge, P.A./Harvey, F.D., eds, *Hellenistic and Roman Sparta. A Tale of Two Cities*, London/New York, 208–32.

Foxhall, L. (1989), 'Household, gender and property in Classical Athens', *Classical Quarterly* 39, 22–44.

Foxhall, L./Forbes, H.A. (1982), 'Sitometreia: the role of grain as a staple food in classical antiquity', *Chiron* 12, 41–90.

Friedl, E. (1962), *Vasilika, A Village in Modern Greece*, New York.

Frisk, H. (1960), *Griechisches etymologisches Wörterbuch*, 2 vols, Heidelberg.

Fuks, A. (1951), '*Kolonòs místhios*: Labour exchange in Classical Athens', *Eranos* 49, 171–3.

Gabrielsen, V. (1986), '*Fanera* and *afanes ousia* in Classical Athens', *Classica et Mediaevalia* 37, 99–114.

—— (1991), *The Athenian Trierarchy*, Odense.
Gallant, T.W. (1982), 'Agricultural systems, land tenure, and the reforms of Solon', *BSA* 77, 111–24.
Gansiniec, Z. (1956), 'Cereals in early archaic Greece', *Archeologia* 8, 1–48 (Polish with a summary in English).
Gardiner, E.N. (1912), 'Panathenaic Amphorae', *JHS* 32, 179–93.
Garnsey, P. (1985), 'Grain for Athens', in Cartledge, P.A./Harvey, F.D., eds, *Hellenistic and Roman Sparta. A Tale of Two Cities*, London/New York, 62–75.
—— (1988a), *Famine and Food Supply in the Graeco-Roman World. Responses to Risk and Crisis*, Cambridge.
—— (1988b), 'Mountain economies in southern Europe', in Whittaker, C.R., ed., *Pastoral Economies in Classical Antiquity* (Cambridge Philological Society Suppl. 14), Cambridge, 196–209.
Gauthier, Ph. (1966), 'Les clérouques de Lesbos et la colonisation athénienne aux Ve siècle', *REG* 79, 64–88.
—— (1973), 'A propos des clérouquies athéniennes du Ve siècle', in Finley, M.I., ed., *Problèmes de la terre en Grèce ancienne*, Paris, 163–78.
Gavrielides, N.E. (1976), 'A Study in the Cultural Ecology of an Olive-Growing Community: The Southern Argolid, Greece', diss. Indiana.
Geographical Handbook Series (1944), 3 vols, The British Admiralty, Naval Intelligence Division edn.
Georgoudi, S. (1974), 'Quelques problèmes de la transhumance dans la Grèce ancienne', *REG* 87, 155–85.
Giardina, A./Schiavone, A. (1981), eds, *L'Italia: Insediamenti e forme economiche*, vol. 1 of *Società romana e produzione schiavistica* (1981), 3 vols, Bari.
Glotz, G. (1920), *Le Travail dans la Grèce ancienne*, Paris.
—— (1923), 'Un transport de marbre pour le portique d'Eleusis (333/2)', *REG* 36, 22–45.
—— (1926), *Ancient Greece at Work*, English transl. by M.R. Dobie, London.
Gluskina, L.M. (1974), 'Studien zu den sozial-ökonomischen Verhältnissen in Attika im 4.Jh.v.u.Z', *Eirene* 12, 111–38.
Gofas, D.C. (1969), 'Les carpologues de Thasos', *BCH* 93, 337–70.
—— 'L'*orkos neidies* à Thasos', *BCH* 95, 245–57.
Grace, V. (1979), *Amphoras and the Ancient Winetrade* (Agora Picture Books 6), Athens. (First edn, 1961.)
Graham, A.J. (1982), in *Cambridge Ancient History*³ III, Part 3, 83–195.
Grassl, H. (1985a), 'Zur Geschichte des Viehhandels im klassischen Griechenland', *Münstersche Beiträge zur Antiken Handelsgeschichte* 4, 77–88.
—— (1985b), 'Hirtenkultur in Griechenland', in *Bericht über den sechzenten österreichischen Historikertag in Krems/Donau, 3.–7. September 1984*, 77–85.
Greenfield, H.J. (1988), 'The origins of milk and wool production in the old world', *Current Anthropology* 29, 573–93.
Gschnitzer, F. (1964), *Studien zur griechischen Terminologie der Sklaverei* I (Akadem. der W. und der Litt. in Mainz. Abh. der Geistes-und Sozialw. Klasse 1963, 13), Wiesbaden.
Guarducci, M. (1935–50), ed., *Inscriptiones Graecae*, 4 vols, Rome.
—— (1937), 'I pascoli del santuario di Alea a Tegea', *Rivista di Filologia e di Istruzione Classica* 65, 169–72.
—— (1952), 'La legge dei tegeati intorno ai pascoli di Alea', *Rivista di Filologia e di Istruzione Classica* 80, 49–68.
Guiraud, P. (1893), *La Propriété foncière en Grèce jusqu'à la conquête romaine*, Paris.

Halstead, P. (1981), 'Counting sheep in neolithic and bronze age Greece', in Hodder, I./Isaac, G./Hammond, N., eds, *Pattern of the Past. Studies in Honour of David Clarke*, Cambridge, 307–39.
—— (1987), 'Traditional and ancient rural economy in mediterranean Europe: plus ça change?', *JHS* 107, 77–87.
—— (1989), 'Agrarian ecology in the Greek Islands: time stress, scale and risk', *JHS* 109, 41–55.
Hammond, N.G.L. (1963), 'The physical geography of Greece and the Aegean', in Wace, A.J.B./Stubbings, F.H., eds, *A Companion to Homer*, London/New York, 269–82.
Hands, A.R. (1968), *Charities and Social Aid in Greece and Rome*, London/Southampton.
Hannestad, L. (1988),'The Athenian potter and the home market', in Christiansen, J./Melander, T., eds, *Proceedings of the Third Symposium on Ancient Greek and Related Pottery*, Copenhagen, 222–30.
Hansen, M.H. (1973), *Atimistraffen i Athen i klassisk tid*, Copenhagen.
—— (1976), *Apagoge, Endeixis and Ephegesis against Kakourgoi, Atimoi and Pheugontes*, Odense.
—— (1977–81), *Det athenske demokrati i 4. århundrede f.Kr.*, 6 vols, Copenhagen.
—— (1978), '*Nomos* and *psefisma* in fourth-century Athens', *GRBS* 19, 315–30 (= Hansen, M.H. (1983), *The Athenian Ecclesia. A Collection of Articles 1976–83*, Copenhagen, 161–76).
—— (1983), *The Athenian Ecclesia. A Collection of Articles 1976–83*, Copenhagen.
—— (1988), *Three Studies in Athenian Demography* (*HFM* 56).
—— (1989a), *Was Athens a Democracy?* (*HFM* 59).
—— (1989b), *The Athenian Ecclesia II. A Collection of Articles 1983–89*, Copenhagen.
—— (1991) *The Athenian Democracy in the Age of Demosthenes*, Oxford.
Hanson, V.D. (1983), *Warfare and Agriculture in Classical Greece* (Bibliotheca di Studi Antichi 40), Pisa.
Harrison, A.R.W. (1968), *The Law of Athens, 1. The Family and Property*, Oxford.
Harrison, J.E. (1903), 'Mystica Vannus Iacchi', *JHS* 23, 292–324.
—— (1904), 'Mystica Vannus Iacchi', *JHS* 24, 241–54.
Hartmann, H.T./Kester, D.E. (1968), *Plant Propagation. Principles and Practices*, New Jersey.
Haussoullier, B. (1879), 'Inscriptions de Chio', *BCH* 3, 230–55.
Hegyi, D. (1976), 'Temene hiera kai temene demosia', *Oikumene* 1, 77–87.
Hehn, V. (1963), *Kulturpflanzen und Haustiere in ihrem Übergang aus Asien nach Griechenland und Italien sowie in das übrige Europa*, Darmstadt. (First printing, 1870.)
Heichelheim, F. (1935), 'Sitos', *RE*, Suppl. 6, 819–92.
Heitland, W.E. (1921), *Agricola*, Cambridge.
Hennig, D. (1985), 'Die "heiligen Häuser" von Delos', *Chiron* 15, 165–86.
—— (1987), 'Kaufverträge über Häuser und Ländereien aus der Chalkidike und Amphipolis', *Chiron* 17, 143–69.
Hentz, G. (1979), 'Les sources grecques dans les écrits des agronomes latins', *Ktema* 4, 151–60.
Herfst, P. (1922), *Le Travail de la femme dans la Grèce ancienne*, Utrecht.
Hermann, J. (1958), *Studien zur Bodenpacht im Recht der Graeco-Aegyptischen Papyri* (Münchener Beiträge zur Papyrusforschung und antiken Rechtsgeschichte 41), Munich.
Hodkinson, S. (1983), 'Social order and the conflict of values in Classical Sparta', *Chiron* 13, 239–81.

—— (1986), 'Land tenure and inheritance in Classical Sparta', *Classical Quarterly* 36, 378–406.

—— (1988), 'Animal husbandry in the Greek polis', in Whittaker, C.R., ed., *Pastoral Economies in Classical Antiquity* (Cambridge Philological Society Suppl. 14), Cambridge, 35–74.

Hodkinson, S./Hodkinson, H. (1981), 'Mantineia and the Mantinike: settlement and society in a Greek polis', *BSA* 76, 239–96.

Holland, L. (1944), 'Colophon', *Hesperia* 13, 91–171.

Hopkins, K. (1967), 'Slavery in Classical Antiquity', in De Reuck, A./Knight, J., eds, *Caste and Race: Comparative Approaches*, London, 166–77.

Hopper, R.J. (1979), *Trade and Industry in Classical Greece*, London.

Hörnschemeyer, A. (1929), *Die Pferdezucht im klassischen Altertum*, diss. Giessen.

Humphreys, S.C. (1967), 'Archaeology and the social and economic history of Classical Greece', *La Parola del Passato* 116, 374–400 (= Humphreys, S.C. (1978), *Anthropology and the Greeks*, London/Henley/Boston, 109–35).

—— (1978), *Anthropology and the Greeks*, London/Henley/Boston.

Isager, S. (1981/2), 'The marriage pattern in Classical Athens. Men and women in Isaios', *Classica et Mediaevalia* 33, 81–96.

—— (1983), *Forpagtning af jord og bygninger i Athen. Attiske indskrifter*, Copenhagen.

—— (1988), 'Once upon a time. On the interpretation of [Aristotle], Oikonomika ii', in Damsgaard-Madsen, A./Christiansen, E./Hallager, E., eds, *Studies in Ancient History and Numismatics Presented to Rudi Thomsen*, Aarhus, 77–83.

—— (1990), 'Som guderne vil: De græske guder som jordejere', *Den Jyske Historiker* 51/2, 33–44.

Isager, S./Hansen, M.H. (1975), *Aspects of Athenian Society in the Fourth Century BC*, Odense.

Jameson, M.H. (1965), 'Notes on the sacrificial calendar from Erchia', *BCH* 89, 154–72.

—— (1976), 'The Southern Argolid: the setting for historical and cultural studies', in Dimen, M./Friedl, E., eds, 74–91.

—— (1977/8), 'Agriculture and slavery in Classical Athens', *Classical Journal* 73, 122–45.

—— (1982), 'The leasing of land in Rhamnous', in *Studies in Attic Epigraphy, History, and Topography, Presented to Eugene Vanderpool* (*Hesperia* Suppl. 19), Princeton, 66–74.

—— (1988), 'Sacrifice and animal husbandry in Classical Greece', in Whittaker, C.R. ed., *Pastoral Economies in Classical Antiquity* (Cambridge Philological Society Suppl. 14), Cambridge, 87–119.

—— (1989), 'Mountains and the Greek city states', in Bergier, J.-F., ed., *Archaeological Aspects of Woodland Ecology*, Oxford, 7–17.

—— 'Agricultural labor in ancient Greece', in Wells, ed.

Jameson, M.H./Runnels, C.N./Van Andel, Tj.H. (forthcoming), eds, *The Southern Argolid. A Greek Countryside from Prehistory to the Present Day*, Stanford.

Jardé, A. (1979), *Les Céreales dans l'antiquité grecque*, Paris. (First edn, 1925.)

—— (1957), 'Vinum', in *Daremberg e Saglio* s.v.

Jasny, N. (1941–2), 'Competition among grains in Classical antiquity', *AHR* 47, 747–64.

—— (1944), *The Wheats of Classical Antiquity*, Baltimore.

Jeppesen, K. (1987), *The Theory of the Alternative Erechtheion*, Aarhus.

Johansen, F. (1982), 'Græske geometriske bronzer', *Meddelelser fra Ny Carlsberg*

Glyptotheket 38, 73–98.
Johnston, A.W./Jones, R.E. (1978), 'The SOS amphora', *BSA* 73, 103–41.
Jones, J.E. (1975), 'Town and country houses of Attica in Classical times', in Mussche, M./Spitaels, P./Goemare-De Poerck, F., eds, *Thorikos and Laurion in Archaic and Classical Times (Miscellanea Graeca* I), Ghent, 63–140.
Jones, J.E./Graham, A.J./Sackett, L.H. (1973), 'An Attic country house below the Cave of Pan at Vari', *BSA* 68, 355–452.
Jones, J.E./Sackett, L.H./Graham, A.J. (1962), 'The Dema House in Attica', *BSA* 57, 75–114.
Jones, N.F. (1987), *Public Organization in Ancient Greece: A Documentary Study*, Philadelphia.
Jongman, W. (1988), 'Adding it up', in Whittaker, C.R., ed., *Pastoral Economies in Classical Antiquity* (Cambridge Philological Society Suppl. 14), Cambridge.
Jost, M. (1985), *Sanctuaires et cultes d'Arcadie* (Etudes Péloponnésiennes ix), Paris.
Kahrstedt, U. (1953), 'Delphoi und das heilige Land des Apollon', in Mylonas, G.E./Raymond, D., eds, *Studies Presented to D.M. Robinson* ii, St Louis, 749–57.
Kallifatides, Th. (1983), *Brennvin och Rosor* [Brandy and Roses], Copenhagen.
Kaltsas, N. (1985), 'He archaïke oikia sto Kopanáki tes Messenias', *Archaiologike Ephemeris*, 207–37.
Kamps, W. (1937), 'Les origines de la fondation cultuelle dans la Grèce ancienne', *Arch. Hist. des Dr. Orient.* 1, 145–79.
Keller, D.R./Rupp, D.W. (1983), eds, *Archaeological Survey in the Mediterranean Area* (BAR 155).
Keller, O. (1909), *Die antike Tierwelt*, 2 vols, Leipzig.
Kent, J.H. (1948), 'The temple estates of Delos, Rheneia and Mykonos', *Hesperia* 17, 243–338.
Kirsten, E. (1956), *Die griechische Polis als historisch-geographisches Problem des Mittelmeerraumes*, Bonn.
Klingenberg, E. (1976), *Platonos nomoi georgikoi und das positive griechische Recht* (Münchener Universitätsschriften, Jur.F., Abh. zur Rechtswissenschaftlichen Grundl.forsch. 17), Berlin.
Knoepfler, D. (1988), ed., *Comptes et inventaire dans la cité grecque. Actes du colloque international d'épigraphie tenu à Neuchâtel du 23 au 26 septembre 1986 en l'honneur de Jacques Tréheux*, Geneva.
Koch (1901), 'Dekate', *RE* 8, 2423–4.
Koerner, R. (1987), 'Zur Landaufteilung in griechischen Poleis in älterer Zeit', *Klio* 69, 443–9.
Kornemann, E. (1963a), 'Bauernstand', *RE*, Suppl. 4, 83–108.
—— (1963b), 'Domänen', *RE*, Suppl. 4, 227–68.
Koster, H.A. (1976), 'The thousand year road', *Expedition* 19.1, 19–28.
Kothe, H. (1975), 'Der Hesiodpflug', *Philologus* 119, 1–26.
Kouremenos, K.E. (1985), *The Sarakatsani. Greek Traditional Architecture*, Athens.
Kozloff, A.P. (1981), *Animals in Ancient Art: From the Leon Mildenberg Collection*, Indiana.
Kozloff, A.P./Mitten, D.G./Sguaitamatti, M. (1986), *More Animals in Ancient Art*, Mainz.
Kraemer, H. (1940), 'Rind', *RE*, Suppl. 7, 1155–85.
Kreissig, H. (1985), 'Zur metaxy-Problematik im Altertum. Einführung in die Diskussion', in Kreissig, H./Kühnert, F., eds, *Antike Abhängigkeitsformen in den griechischen Gebieten ohne Polisstruktur und den römischen Provinzen* (Schriften zur Geschichte und Kultur der Antike 25), Berlin, 9–11.

Kreissig, H/Kühnert, F. (1985), eds, *Antike Abhängigkeitsformen in den griechischen Gebieten ohne Polisstruktur und den römischen Provinzen* (Schriften zur Geschichte und Kultur der Antike 25), Berlin.
Lacroix, M. (1916), 'Une famille de Délos', *REG* 29, 188–237.
Lambrinudakis, V. (1982), 'Antike Niederlassungen auf dem Berge Aipos von Chios', in Papenfuss, D./Strocka, M., eds, *Palast und Hütte*, Mainz am Rhein, 375–94.
Lambrinoudakis, V. (1986), 'Ancient farmhouses on Mount Aipos', in Boardman, J./Vaphopoulou-Richardson, C.E., eds, *Chios: A Conference at the Homereion in Chios 1984*, Oxford, 295–304.
Langdon, M.K. (1985), 'The territorial basis of the Attic demes', *Symbolae Osloenses* 60, 5–15.
—— (1987), 'An Athenian decree concerning Oropos', *Hesperia* 56, 47–58.
Langdon, M.K./Watrous, L.V. (1977), 'The farm of Timesios: rock-cut inscriptions in South Attica', *Hesperia* 46, 162–77.
Latte, K. (1933), 'Moria', *RE* 16.1, 302–3.
Laurenti, R. (1968), *Studi sull'Economico attribuito ad Aristotele*, Milan.
Lauter, H. (1980), 'Zu Heimstätten und Gutshäusern im klassischen Attika', in Krinzinger, F., ed., *Forschungen und Funde: Festschrift Bernhard Neutsch* (Innsbr. Beitr. zur Kulturwiss. 21), Innsbruck, 279–86.
—— (1981), 'Klassisches Landleben in Attika', *Hellenika, Jahrbuch für die Freunde Griechenlands*, 162–4.
Lefebvre des Noëttes, R. (1931), *L'Attelage, le cheval de selle à travers les âges*, Paris.
Lepore, E. (1968), 'Per una fenomenologia storica del rapporto città-territorio in Magna Grecia', in *La città e il suo territorio. Atti del settimo convegno di studi sulla Magna Grecia, Taranto 8–12 ottobre 1967*, Naples, 29–66.
—— (1973), 'Problemi dell'organizzazione della chora coloniale', in Finley, M.I., ed., *Problèmes de la terre en Grèce ancienne*, Paris, 15–47.
Leveau, Ph. (1985), ed., *L'Origine des richesses dépensées dans la ville antique*, Aix-en-Provence.
Lewis, D.M. (1959a), 'Attic manumissions', *Hesperia* 28, 208–38 (Greek text reprinted (1962) in *SEG*, 18, 36).
—— (1959b), 'Law on the Lesser Panathenaia', *Hesperia* 28, 239–47.
—— (1968), 'Dedications of phialai at Athens', *Hesperia* 37, 368–80 (Greek text reprinted (1971) in *SEG*, 25, 178–80).
—— (1973), 'The Athenian Rationes Centesimarum', in Finley, M.I., ed., *Problèmes de la terre en Grèce ancienne*, Paris, 187–212.
Linders, T. (1975), *The Treasurers of the Other Gods in Athens and their Functions* (Beiträge zur klassischen Philologie 62), Meisenheim am Glan.
—— (1987), 'Gods, gifts, society', in Linders, T./Nordquist, G., eds, *Gifts to the Gods. Proceedings of the Uppsala Symposium 1985* (Boreas 15), Uppsala, 115–22.
—— (1988), 'The purpose of inventories: a close reading of the Delian inventories of the independence', in Knoepfler, D., ed., 37–47.
Linders, T./Nordquist, G. (1987), eds, *Gifts to the Gods. Proceedings of the Uppsala Symposium 1985* (Boreas 15), Uppsala.
Lohmann, H. (1985), 'Landleben im klassischen Attika. Ergebnisse und Probleme einer archäologischen Landesaufnahme des Demos Atene', *Jahrbuch Ruhr-Universität Bochum*, 71–96.
—— 'Agriculture and country life in Classical Athens', in Wells, ed.
Loreaux, N. (1981a), *L'Invention d'Athènes*, Paris.
—— (1981b), *Les Enfants d'Athéna: idées athéniennes sur la citoyenneté et la division*

des sexes, Paris.
—— (1981c), 'La cité comme cuisine et comme partage', *Annales (ESC)* 36, 614–22.
Lotze, D. (1985), 'Zu neuen Vermutungen über Landleute im alten Sikyon', in Kreissig, H./Kühnert, F., eds, *Antike Abhängigkeitsformen in den griechischen Gebieten ohne Polisstruktur und den römischen Provinzen* (Schriften zur Geschichte und Kultur der Antike 25), Berlin, 20–8.
McDonald, W.A./Rapp, G.R. (1972), eds, *The Minnesota Messenia Expedition. Reconstructing a Bronze Age Environment*, St Paul/Oxford/Toronto.
MacDowell, D.M. (1986), *Spartan Law*, Edinburgh.
—— (1989), 'The *oikos* in Athenian law', *Classical Quarterly* 39, 10–21.
McNall, S.G. (1974), *The Greek Peasant*, Washington.
Malkin, I. (1987), *Religion and Colonization in Ancient Greece*, Leiden/New York/Copenhagen/Cologne.
Manzoufas, G. (1967), 'La loi thasienne *Gleukos mede oinon* sur le commerce du vin', Athens.
Martin, R. (1973), 'Rapports entre les structures urbaines et les modes de division et d'exploitation du territoire', in Finley, M.I., ed., *Problèmes de la terre en Grèce ancienne*, Paris, 97–112.
—— (1974a), *L'Urbanisme dans la Grèce antique*, Paris. (2nd revised edn.)
—— (1974b), 'Problèmes d'urbanisme dans les cités grecques de Sicile', *Kokalos* 18–19, 348–65.
Meiggs, R. (1982), *Trees and Timber in the Ancient Mediterranean World*, Oxford.
Meiggs, R./Lewis, D. (1969), eds, *A Selection of Greek Historical Inscriptions to the End of the Fifth Century B.C.*
Métraux, G.P.R. (1978), *Western Greek Land-Use and City-Planning in the Archaic Period*, New York/London.
Metzger, R.R. (1973), *Untersuchungen zum Haftungs-und Vermögensrecht von Gortyn*, Basel.
Michell, H. (1963), *The Economics of Ancient Greece*, Cambridge. (First edn, 1940.)
Mickwitz, G. (1937), 'Economic rationalism in Greco-Roman agriculture', *English History Review* 208, 577–89.
Mikalson, J.D. (1975), *The Sacred and Civil Calendar of the Athenian Year*, Princeton/London.
—— (1977), 'Religion in the Attic Demes', *AJPh* 98, 424–35.
—— (1983), *Athenian Popular Religion*, Chapel Hill/London.
Millet, P. (1982), 'The Attic *Horoi* reconsidered in the light of recent discoveries', *Opus* I, 219–49.
Moggi, M. (1981), 'Alcuni episodi della colonizzazione ateniese', in Cataldi, S./Moggi, M./Nenci, G./Panessa, G., *Studi sui rapporti interstatali nel mondo antico*, Pisa, 1–55.
—— (1987), 'Organizzazione della *chora*, proprietà fondiaria e *homonoia*: il caso di Turi', *Annali della Scuola Superiore di Pisa* 17, 65–88.
Molinier, S. (1914), *Les Maisons sacrées de Delos au temps de l'indépendance de l'île, 315–166/5 av. J.-C.*', (Bibliothèque de la Faculté des Lettres 31), Paris.
Moritz, L.A. (1949), '*Alfita* – a Note', *Classical Quarterly* 43, 113–17.
—— (1955a), 'Husked and "Naked" Grain', *Classical Quarterly* 5, 129–34.
—— (1955b), 'Corn', *Classical Quarterly* 5, 135–41.
—— (1958), *Grain-Mills and Flour in Classical Antiquity*, Oxford.
Mossé, C. (1977), 'Les périèques lacédémoniens. A propos d'Isocrate, *Panathénaïque* 177 sqq.', *Ktema* 2, 121–4.
Müller, R. (1985), 'Polis und Ethnos, Sklaverei und andere Abhängigkeitsformen in der griechischen Gesellschaftstheorie', in Kreissig, H./Kühnert, F., eds, *Antike*

Abhängigkeitsformen in den griechischen Gebieten ohne Polisstruktur und den römischen Provinzen (Schriften zur Geschichte und Kultur der Antike 25), Berlin, 49–55.
Nemes, Z. (1980), 'The public property of demes in Attica', *Acta Classica Univ. Scient. Deprecen.* 16, 3–8.
Nilsson, M.P. (1920), *Primitive Time-Reckoning*, Lund.
—— (1930), *Solkalender og Solreligion*, Copenhagen/Oslo.
—— (1951), 'Zeitrechnung, Apollo und der Orient', *Opuscula Selecta* i (Acta Inst. Athen. Regni Sueciae II. 1), Lund, 36–61.
—— (1955), 'Das frühe Griechenland von innen gesehen', *Historia* 3, 257–82.
Nollé, J. (1983), 'Zum Landbau von Side (d'après les inscriptions)', *Epigraphica Anatolica* 1, 119–29.
Noonan, T.S. (1973), 'The grain trade of the Black Sea in antiquity', *AJPh* 94, 231–42.
Nussbaum, G. (1960), 'Labour and status in the Works and Days', *Classical Quarterly* 54, 213–20.
Ober, J. (1981), 'Rock-cut inscriptions from Mount Hymettos', *Hesperia* 50, 68–77.
OEEC Report 56 (1951), *Pasture and Fodder Development in Mediterranean Countries*.
Olck, F. (1893), 'Ackerbau', *RE* 1.1, 261–86.
—— (1897a), 'Biene', *RE* 3, 431–50.
—— (1897b), 'Bienenzucht', *RE* 3, 450–7.
—— (1907), 'Esel', *RE* 6/1, 625–76.
Oliva, P. (1971), *Sparta and her Social Problems*, Prague.
Orth, E. (1910),'Geflügelzucht', *RE* 7/1, 903–27.
—— (1913), 'Huhn', *RE* 8/2, 2519–36.
—— (1921a), 'Schaf', *RE* II 2A/1, 373–99.
—— (1921b), 'Schwein', *RE* 2A/1, 801–5.
—— (1924), 'Landwirtschaft', *RE* 12, 624–76.
—— (1929), 'Stier', *RE* 3A/2, 2495–520.
Osborne, R. (1985a), *Demos: The Discovery of Classical Attika*, Cambridge.
—— (1985b), 'Buildings and residence on the land in Classical and Hellenistic Greece: the contribution of epigraphy', *BSA* 80, 119–28.
—— (1985c), 'The land-leases from Hellenistic Thespiai: a reexamination', in *La Béotie antique*, Paris, 317–23.
—— (1987), *Classical Landscape with Figures: The Ancient Greek City and its Countryside*, London.
—— (1988), 'Social and economic implications of the leasing of land and property in Classical and Hellenistic Greece', *Chiron* 18, 279–323.
—— (1991), 'Pride and prejudice, sense and subsistence: exchange and society in the Greek city', in Rich J./Wallace-Hadrill, eds, *City and Country in the Ancient World*, London/New York, 119–45.
Owens, E.J. (1983), 'The *koprologoi* at Athens in the fifth and fourth centuries BC', *Classical Quarterly* 33, 44–50.
Papenfuss, D./Strocka, V.M. (1982), eds, *Palast und Hütte*, Mainz am Rhein.
Parke, H.W. (1977), *Festivals of the Athenians*, London.
Parker, A.J. (1984), 'Shipwrecks and ancient trade in the Mediterranean', *Archaeological Review from Cambridge* 3.2, 99–112.
Parker, R. (1983), *Miasma. Pollution and Purification in Early Greek Religion*, Oxford.
—— (1987), 'Festivals of the Attic demes', in Linders, T./Nordquist, G., eds, *Gifts to the Gods. Proceedings of the Uppsala Symposium 1985* (*Boreas* 15), Uppsala,

137–47.
Pease, A.S. (1937a), 'Oleum', *RE* 17.2, 2454–74.
—— (1937b), 'Olbaum', *RE* 17.2, 1998–2022.
Pečírka, J. (1970), 'The polis of Chersonesos in the Crimea', in Rosa, L. de, ed., *Ricerche storici ed economiche in memoria di Corrado Barbagallo*, Naples, i, 459–77.
—— (1971), 'Die Landgüter der Milesier', *Jahrbuch für Wirtschaftsgeschichte* 2 (*Festschrift Welskopf*), 55–61.
—— (1973), 'Homestead farms in Classical and Hellenistic Hellas', in Finley, M.I., ed., *Problèmes de la terre en Grèce ancienne*, Paris, 113–47.
Peek, W. (1942), 'Attische Inschriften', *AM* 67, 1–217.
Pfuhl, E. (1923), *Malerei und Zeichnung der Griechen*, Munich.
Philippson, A. (1948), *Das Klima Griechenlands*, Bonn.
—— (1950–9), *Die griechischen Landschaften*, 4 vols, Frankfurt am Main.
Piérart, M. (1985), 'Athènes et Milet' II, *Museum Helveticum* 42, 276–99.
Pippidi, D.M. (1973), 'Le problème de la main-d'œuvre agricole dans les colonies grecques de la Mer Noire', in Finley, M.I., ed., *Problèmes de la terre en Grèce ancienne*, Paris, 63–82.
Pleket, H.W. (1973), 'Economic history of the ancient world and epigraphy: some introductory remarks', in *Akten des vi. Intern. Kongr. für Griech. und Lat. Epigraphie, München 1972 (Vestigia 17)*, Munich, 243–57.
de Polignac, F. (1984), *La Naissance de la cité grecque: cultes, éspace et société*, Paris.
Pomtow, H. (1906), 'Eine delphische *stasis* im Jahre 363 v.Chr.', *Klio* 6, 89–126.
Pouilloux, J. (1954), *Recherches sur l'histoire et les cultes de Thasos* (Etudes thasiennes 3), Paris.
Preaux, C. (1973), *La Lune dans la pensée grecque*, Brussels.
Preisigke, F. (1919), 'Die Begriffe PYRGOS und STEGE bei der Hausanlage', *Hermes* 54, 423–32.
Prickett, J.L. (1980), 'A Scientific and Technological Study of Topics Associated with the Grape in Greek and Roman Antiquity', diss. University of Kentucky.
Pritchett, W.K. (1953), 'The Attic Stelai, Part 1', *Hesperia* 22, 225–99.
—— (1956), 'The Attic Stelai, Part 2', *Hesperia* 25, 178–317.
—— (1960), *Marathon* (University of California Publications in Classical Archaeology 4.2), Berkeley/Los Angeles.
—— (1963), *Ancient Athenian Calendars on Stone* (University of California Publications in Classical Archaeology 4.4), Berkeley/Los Angeles.
—— (1965), *Studies in Ancient Greek Topography*. Part I (University of California Publications in Classical Studies 1), Berkeley/Los Angeles.
—— (1969), *Studies in Ancient Greek Topography*. Part II (University of California Publications in Classical Studies 4), Berkeley/Los Angeles.
Psarraki-Belesióte, N. (1978), *Traditional Methods of Cultivation*, Athens.
Rackham, O. (1983), 'Observations on the Historical Ecology of Boeotia', *BSA* 78, 291–351.
Rackham, O./Moody, J.A., 'Terraces', in Wells, ed.
Rankin, E.M. (1907), *The Rôle of the Mageiroi in the Life of the Ancient Greeks*, Chicago.
Redfield, J. (1977/8), 'The women of Sparta', *Classical Journal* 73, 146–61.
Reilly, L.C. (1978), *Slaves in Ancient Greece. Slaves from Greek Manumission Inscriptions*, Chicago.
Renfrew, C./Wagstaff, M. (1982), eds, *An Island Polity: The Archaeology of Exploitation in Melos*, Cambridge.
Renfrew, J. (1973), *Palaeoethnobotany. The Prehistoric Foodplants of the Near East*

and Europe, London.
Rhodes, P.J. (1981), *A Commentary on the Aristotelian Athenaion Politeia*, Oxford.
Rich, J./Wallace-Hadrill, A. (1991), eds, *City and Country in the Ancient World*, London/New York.
Richardson, N.J./Piggott, S. (1982), 'Hesiod's waggon, text and technology', *JHS* 102, 225–9.
Richardson, R.B. (1895), 'A sacrificial calendar from the Epakria', *AJA* 10, 209–26.
Richter, G.M. (1930), *Animals in Greek Sculpture*, Oxford.
Richter, W. (1968), *Die Landwirtschaft im homerischen Zeitalter. Mit einem Beitrag: Landwirtschaftlische Geräte, von Wolfgang Schiering* (Archaeologia Homerica II, H), Göttingen.
—— (1972), 'Ziege', *RE* 2/10/1, 398–433.
Riezler, K. (1907), *Uber Finanzen und Monopole im alten Griechenland. Zur Theorie und Geschichte der antiken Stadtwirtschaft. i, Pseudoaristoteles Okonomik B*, Berlin.
Robert, L. (1949a), 'Épitaphe d'un berger à Thasos', *Hellenica* 7, 152–60.
—— (1949b), 'Les chèvres d'Héracleia', *Hellenica* 7, 161–70.
—— (1960), 'Sur une loi d'Athènes relative aux petites Panathénées', *Hellenica* 11–12, 189–203.
—— (1973), 'Les juges étrangers dans la cité grecque', in Von Caemmerer, E., ed., *Xenion, Festschrift für Pan. J. Zepos*, Athens, i, 765–82.
Rocchi, G.D. (1981), 'Gli insediamenti in villaggi nella Grecia del V e del IV sec. a.C.', *Istituto Lombardo (Memorie Lett.)* 36, 325–86.
Roebuck, C. (1945), 'A note on the Messenian economy and population', *CPh* 40, 149–65.
—— (1986), 'Chios in the sixth century BC', in Boardman, J./Vaphopoulou-Richardson, C.E., eds, 81–8.
Roesch, P. (1973), 'Pouvoir fédéral et vie économique des cités dans la Béotie hellénistique', *Akten des VI. Intern. Kongr. für Griech. und Lat. Epigraphie. München 1972 (Vestigia* 17), Munich, 259–70.
—— (1982), *Etudes Béotiennes*, Paris.
—— (1985), 'Les femmes et la fortune en Béotie', in *La Femme dans le monde méditerranéen. I, Antiquité.* (Travaux de la Maison de l'Orient 10), Lyons, 71–84.
Roobaert, A. (1977), 'Le danger hilote?', *Ktema* 2, 141–55.
Roussel, P. (1916), *Délos, colonie athénienne* (Bibliothèque des Ecoles Françaises d'Athènes et de Rome 111), Paris.
Runnels, C.N. (1981), 'A Diachronic Study and Economic Analysis of Millstones from the Argolid', diss. Indiana University.
Ruschenbusch, E. (1966), *Solonos nomoi* (*Historia*, 9), Wiesbaden.
Salmon, J.B. (1984), *Wealthy Corinth: A History of the City to 338 B.C.*, Oxford.
Salviat, F. (1986), 'Le vin de Thasos. Amphores, vin et sources écrites', in Empereur, J.-Y./Garlan, Y., eds, *Recherches sur les amphores grecques* (*BCH* Suppl. 13), Paris, 145–95.
Salviat, F./Vatin, C. (1974), 'Le cadastre de Larissa', *BCH* 98, 247–62.
Samuel, A.E. (1972), *Greek and Roman Chronology: Calendars and Years in Classical Antiquity*, Munich.
Sarikakis, Th. Ch. (1986), 'Commercial relations between Chios and other Greek cities in antiquity', in Boardman, J./Vaphopoulou-Richardson, C.E., eds, 121–31.
Sartre, M. (1979), 'Aspects économiques et aspects religieux de la frontière dans les cités grecques', *Ktema* 4, 213–24.

Schaps, D.M. (1979), *Economic Rights of Women in Ancient Greece*, Edinburgh.
Scheidel, W. (1990), 'Feldarbeit von Frauen in der antiken Landwirtschaft', *Gymnasium* 97, 405–31.
Schiering, W. (1968), 'Die landwirtschaftlische Geräte', in Richter, W., *Die Landwirtschaft im homerischen Zeitalter* (Archaeologica Homerica II. H), Göttingen, 147–58.
Schnebel, M. (1925), *Die Landwirtschaft im hellenistischen Ägypten* (Münchener Beiträge zur Papyrusforsch. und antiken Rechtsgeschichte 7), Munich.
Seltman, C. (1957), *Wine in the Ancient World*, London.
Semple, E.C. (1922), 'The influence of geographic conditions upon ancient mediterranean stock-raising', *Annals of the Association of American Geographers* 10/11/12, 3–38.
—— (1932), *The Geography of the Mediterranean Region: Its Relation to Ancient History*, London.
Shear, T.L., Jr. (1987), 'Tax tangle, ancient style', *ASCL Newsletter* Spring, 8.
Simon, E. (1965), 'Attische Monatsbilder', *JdI* 80, 105–23.
—— (1983), *Festivals of Attica. An Archaeological Commentary*, Wisconsin.
Skydsgaard, J.E. (1968), *Varro the Scholar. Studies in the First Book of Varro's De Re Rustica (ARID* 4, suppl.), Copenhagen.
—— (1969), 'Nuove ricerche sulla villa rustica Romana fino all'epoca di Traiano', *ARID* 5, 25–40.
—— (1987), 'L'Agricoltura greca e romana: tradizioni e confronto', *ARID* 16, 7–24.
—— (1988a), 'Solon's *tele* and the agrarian history, a note', in Damsgaard-Madsen, A./Christiansen, E./Hallager, E., eds, *Studies in Ancient History and Numismatics Presented to Rudi Thomsen*, Aarhus, 50–4.
—— (1988b), 'Transhumance in ancient Greece', in Whittaker, C.R., ed., *Pastoral Economies in Classical Antiquity* (Cambridge Philological Society Suppl. 14), Cambridge, 75–86.
—— (1990), 'Studiet af det oldgræske landbrug', *Den Jyske Historiker* 51/2, 33–43.
Snodgrass, A. (1977), *Archaeology and the Rise of the Greek State*, Cambridge.
—— (1980), *Early Greece*, London.
—— (1987), *An Archaeology of Greece*, Berkeley/Los Angeles/London.
—— (1991) 'The rural landscape and its political significance', Opus V-VIII, 53–70.
Sokolowski, F. (1954), 'Fees and taxes in the Greek cults', *Harvard Theological Review* 47, 153–64.
Sordi, M. (1984a), 'Il santuario di Olimpia e la guerra d'Elide' in Sordi, M., ed., 20–30.
—— (1984b), ed., *I santuari e la guerra nel mondo classico* (Contributi dell'Istituto di Storia Antica 10), Milan.
Sordinas, A. (1971), *Old Olive Oil Mills and Presses in the Island of Corfu, Greece* (Memphis State Univ., Anthropological Rsc. Center, Occasional Papers 5), Memphis.
Sparkes, B.A. (1962), 'The Greek kitchen. Part 1', *JHS* 82, 121–37.
—— (1965), 'The Greek kitchen. Part 2', *JHS* 85, 162–3.
—— (1976), 'Treading the grapes', *BABesch* 51, 47–63.
Stanier, R.S. (1953), 'The cost of the Parthenon', *JHS* 73, 68–76.
Stanley, P.V. (1980), 'Two Thasian wine laws: a reexamination', *The Ancient World* 3, 88–93.
Starr, Ch.G. (1958), 'An overdose of slavery', *Journal of Economic History* 18, 17–32.
Steensgaard, N. (1973), *Carracks, Caravans and Companies: The Structural Crisis in the European–Asian Trade in the Early Seventeenth Century*, Copenhagen.
Steier, A. (1938), 'Pferd', *RE* 19, 1430–44.

Steiner, G. (1969), 'Farming', in Roebeuck, C., ed., *The Muses at Work*, Cambridge, Mass., 148–70.
Steinhauer, G. (n.d.), *Museum of Sparta*.
Stillwell, R. (1975), ed., *Princeton Encyclopedia of Classical Sites*, Princeton.
Stubbings, F.H. (1963), 'Food and agriculture', in Wace, A.J.B./Stubbings, F.H., eds, *A Companion to Homer*, London/New York, 523–30.
Szegedy-Maszak, A. (1981), *The nomoi of Theophrastus*, New York.
Thiel, J.H. (1925), 'Zu altgriechischen Gebühren', *Klio* 20, 54–67.
Thompson, K. (1963), *Farm Fragmentation in Greece* (Center of Economic Research. Research Monograph Series 5), Athens.
Thomsen, R. (1964), *Eisphora. A Study of Direct Taxation in Ancient Athens*, Copenhagen.
Tigerstedt, N.E. (1965), *The Legend of Sparta in Classical Antiquity* I, Stockholm.
Tiverios, M. (1974), '*Panathenaïka*', *AD* 29, 142–53.
Tod, M.N. (1948), ed., *A Selection of Greek Historical Inscriptions* II, Oxford.
Valavanis, P. (1986), 'Les amphores Panathénaïques et le commerce athénien de l'huile', in Empereur J.-Y./Garlan Y., eds, *Recherches sur les amphores grecques* (*BCH* Suppl. 13), Paris, 453–60.
Vallet, G. (1968), 'La cité et son territoire dans les colonies grecques d'Occident', in *La città et il suo territorio. Atti del Settimo Convegno di Studi sulla Magna Grecia, Taranto 8–12 ottobre 1967*, Naples, 67–142.
Van Andel, Tj.H./Runnels, C.N. (1987), *Beyond the Acropolis. A Rural Greek Past*, Stanford.
Van Andel, Tj.H./Runnels, C.N./Pope, K.O. (1986), 'Five thousand years of land use and abuse in the southern Argolid, Greece', *Hesperia* 55, 103–28.
Vanbremeersch, N. (1987), 'Représentation de la terre et du travail agricole chez Pindare', *Quaderni di Storia* 13, 73–95.
Van Effenterre, H./Van Effenterre, M. (1985), 'Nouvelles lois archaïques de Lyttos', *BCH* 109, 157–88.
Van Groningen, B.A. (1925), 'De rebus Byzantiorum', *Mnemosyne* 53, 211–22.
—— (1933), *Aristote. Le second livre de l'économique. Edité avec une introduction et un commentaire critique et explicatif*, Leyden.
Vatin, C. (1963), 'Le bronze Pappadakis. Etude d'une loi coloniale', *BCH* 87, 1–19.
—— (1976), 'Jardins et services de voirie', *BCH* 100, 555–64.
Vial, Cl. (1984), *Délos indépendante* (*BCH* Suppl. 10), Paris.
Vickery, K.F. (1936), *Food in Early Greece* (Illinois Studies in Social Sciences 20.3), Urbana.
Vigneron, P. (1968), *Le Cheval dans l'antiquité Gréco-Romaine* I–II, Nancy.
Vita-Finzi, C. (1969), *The Mediterranean Valleys. Geological Changes in Historical Times*, Cambridge.
Wace, A.J.B./Stubbings, F.H. (1963), eds, *A Companion to Homer*, London/New York.
Wagstaff, J.M. (1982), *The Development of Rural Settlement. A Study of the Helos Plain in Southern Greece*, Amersham.
Walbank, M.B. (1982a), 'Regulations for an Athenian festival', *Studies in Attic Epigraphy, History and Topography Presented to Eugene Vanderpool* (*Hesperia* Suppl. 19), Princeton, 173–82.
—— (1982b), 'The confiscation and sale by the *poletai* in 402/1 of the property of the Thirty Tyrants', *Hesperia* 51, 74–98.
—— (1983), 'Leases of sacred properties in Attica', Part 1, *Hesperia* 52, 100–35; Parts 2–4, 177–231.
—— (1984), 'Leases of sacred properties in Attica', Part 5, *Hesperia* 53, 361–8.

—— (1985), 'Leases of sacred properties in Attica, a correction', *Hesperia* 54, 140.
Walcot, P. (1970), *Greek Peasants, Ancient and Modern. A Comparison of Social and Moral Values*, Manchester.
Wallace, W. (1947), 'The demes of Eretria', *Hesperia* 16, 115–46.
Watrous, L.V. (1982), 'An Attic farm near Laurion', in *Studies in Attic Epigraphy, History and Topography. Presented to Eugene Vanderpool* (*Hesperia* Suppl. 19), Princeton, 193–8.
Wehrli, Cl. (1970), 'Les Karpologoi', *Museum Helveticum* 27, 104–6.
Wells, B. (1992), ed., *Agriculture in Ancient Greece. Proceedings of the Seventh International Symposium at the Swedish Institute at Athens 16–17 May 1990*, Stockholm.
West, M.L. (1978), *Hesiod, Works and Days*, Oxford.
White, K.D. (1967a), 'Latifundia', *BICS* 14, 68–79.
—— (1967b), *Agricultural Implements of the Roman World*, Cambridge.
—— (1971–2), *The Great Chesterfield Scythes* (*Proceedings of the Hungarian Agricultural Museum*), Budapest.
—— (1975), *Farm Equipment of the Roman World*, London.
—— (1977), *Country Life in Classical Times*, London.
Whitehead, D. (1986), *The Demes of Attica 508/7–ca.250 BC*, Princeton.
Whittaker, C.R. (1988), ed., *Pastoral Economies in Classical Antiquity* (Cambridge Philological Society Suppl. 14), Cambridge.
Wilkinson, T.J. (1982), 'The definition of ancient manured zones by means of extensive sherd-sampling techniques', *Journal of Field Archaeology* 9, 323–33.
—— (1989), 'Extensive sherd scatters and land-use intensity: some recent results', *Journal of Field Archaeology* 16, 31–46.
Will, E. (1972), *Le Monde grec et L'Orient. I, Le Ve siècle*, Paris.
Willetts, R.F. (1967), *The Law Code of Gortyn* (*Kadmos* Suppl. 1), Berlin.
Winkelstern, K. (1933), *Die Schweinezucht im klassischen Altertum*, diss. Giessen.
Wolf, E. (1966), *Peasants*, New Jersey.
Wood, E.M. (1983), 'Agricultural slavery in Classical Athens', *American Journal of Ancient History* 8, 1–47.
—— (1988), *Peasant-Citizen and Slave*, London/New York.
Yalouris, E. (1986), 'Notes on the topography of Chios', in Boardman, J./Vaphopoulou-Richardson, C.E., eds, 141–68.
Yanuchevitch, Z./Nikolayenko, G./Kuzminova, N. (1985), 'La viticulture à Chersonèse de Taurique aux IVe–IIe siècles av. N.E. d'après les recherches archéologiques et paléoethnobotaniques', *Revue Archéologique* I, 115–22.
Young, J.H. (1956a), 'Ancient towers on the island of Siphnos', *AJA* 60, 51–5.
—— (1956b), 'Studies in South Attica: country estates at Sounion', *Hesperia* 25, 122–46.
Zeissig, K. (1934), *Die Rinderzucht im alten Griechenland*, diss. Giessen.
Ziehen, L. (1939), 'Opfer', *RE* 18.1, 579–627.

INDEX OF PASSAGES CITED

Aelian
  De natura Animalium 16.32: 103
Aischines
  1.94–105: 145; 3.21: 184; 3.107–13: 192; 3.108–12: 186
Andokides
  1.133: 140; fr.3: 102
Aristophanes
  Acharnians 736 ff.: 103
  Birds 709–15: 161
  Peace 925–1126: 175
Anthology 6.104: 56
Aristotle
  Athenaion Politeia; 16.4: 143, 145; 47.3–4: 181; 60: 204–5
  Historia Animalium 84; 488a: 85; 521b ff.: 91; 522b: 111; 542a: 85; 553a ff.: 96; 571 ff.: 85; 573a: 93; 573b: 91; 574a: 96; 575a: 89; 575b: 86; 577a ff.: 87; 577b ff: 87; 595a: 93; 595b: 87; 599a: 85; 610b: 91; 632a: 89
  Nicomachean Ethics 8.9.5: 160
  Politics 118, 127, 130, 133; 1259a: 66; 1267b–1268b: 122; 1269a–b: 139, 152; 1269b: 131; 1270a 23–5: 131; 1270b: 132; 1271a: 139; 1271b: 142; 1272a: 123, 139; 1276b2: 120; 1303a: 125; 1319a: 125; 1320a: 137; 1328b–1329a: 150; 1329a: 153; 1329b–1330a: 122; 1335b20–27: 126; 1365a39–b14: 126
  fr. 538: 152
  fr. 586: 152
Pseudo-Aristotle
  Oeconomica 119; 1347a: 136, 141; 1347b: 143; 1348b–49a: 146; 1350a: 142
  Rhetorica ad Alexandrum 1425: 123
Athenaeus
  Deipnosophists; 141: 138; 143b: 139

Bassus, Cassianus
  Geoponica 38
Behrend 1970
  no.29: 188
  no.30: 190

Cato
  De Agricultura 60: 105
Columella
  5.9.16: 39; 6.1.3: 96

Demosthenes
  12.21: 170; 14.16: 141; 14.27: 136, 140; 32–5: 118; 42: 78; 42.7: 106; 43.71: 204; 47.52: 101; 53.15: 36; 55: 118; 55.19: 148; 55.23–4: 147; 56: 118; 57.32: 145; 59.27: 140
Dio Chrysostomos
  7.42: 56, 100; 36.11: 201
Diodorus Siculus
  11.72.2: 175; 12.11.1–2: 125; 16.83.2: 175
Dionysios of Halicarnassos
  Argumentum ad Lysiam 34: 79
  Hellenica Oxyrhynchia 8.3: 100

FGH
(Jacoby, F. (1923–), ed., *Die Fragmente der griechischen Historiker*, Berlin and Leyden) 399: 139

INDEX OF PASSAGES CITED

GD
(H. Collitz *et al.* (1884–1915), eds, *Sammlung der griechischen Dialekt-Inschriften*, Göttingen); 2.2501 4.15–26: 192

Harpocration
s.v. *eranizontes*: 145

Herodotus
1.89: 173; 2.135: 173; 3.55: 173; 3.58: 70; 5.77: 173; 5.82: 203; 6.46: 139; 7.132: 173; 8.55: 203; 9.81: 173

Hesiod 20
Shield 299: 32; 301: 61
Theogony 22: 84; 971: 24
Works and Days 173: 21; 382 ff.: 7, 22, 161; 405: 84; 423 ff.: 55; 427 ff.: 46; 430: 67; 436: 84; 437: 84; 462 ff.: 22; 465: 163; 470: 52; 473: 52; 485 ff.: 24; 493: 67; 504: 164; 515 ff.: 84; 564 ff.: 26; 571 ff.: 25; 590: 84; 597 ff.: 25; 606: 84, 109; 609 ff.: 26; 612–14: 56; 786: 84

Hippocrates
De Natura Puerorum 29: 94

Homer
Iliad 2.469 ff.: 97; 5.136 ff.: 98; 5.499 ff.: 25, 55; 5.554 ff.: 97; 5.902 f.: 97; 6.236: 97; 10.351: 22, 47; 11.67 ff.: 25; 11.172: 97; 11.548 ff.: 97; 11.678 ff.: 97; 12.451 ff.: 97; 13.702 ff.: 22, 47; 13.588: 42, 55; 15.630 ff.: 98; 16.352: 97; 17.53 ff.: 22, 35; 18.520 ff.: 98; 18.541 ff.: 22, 47; 18.550 ff.: 25, 52; 18.561 ff.: 26; 18.575 ff.: 98; 20.495 ff.: 25; 21.79: 97; 21.257 ff.: 52, 112; 23.834 ff.: 46

Odyssey 1.431: 97; 4.604: 21; 5.127: 22; 6.4 ff.: 9, 70; 7.112 ff.: 26; 7.115 ff.: 41; 7.123–5: 56; 7.125: 61; 9.106 ff.: 27; 9.110: 21; 9.246 ff.: 91; 10.411: 98; 11.588 ff.: 41; 12.127 ff.: 96; 13.31: 22, 47; 14.100 ff.: 97; 17.297: 111; 18.357: 81; 18.366 ff.: 52; 18.371 ff.: 47; 19.112: 21; 19.536 ff.: 95; 24.224: 81; 24.226: 27

IC
(M. Guarducci (1935–50), ed., *Inscriptiones Creticae*, 4 vols, Rome)
I 18: 139
IV 11: 163; 42b: 123; 43b: 147; 72: 150; 73a: 148; 77: 139; 184: 139

ID
(Durrbach, F. *et al.* (1926–), eds, *Inscriptions de Delos*, Paris); 366: 193; 442: 195; 503: 194–6

IG
(*Inscriptiones Graecae* (1873–), Berlin)
I^2 6: 171; 94: 36; 313–14: 172
I^3 6: 171; 71: 171; 78: 170–1; 101: 173; 130: 173; 386: 172
II/III^2 140: 173; 204: 185; 334: 178, 188; 1496: 179; 1553–9: 154; 1638–40: 193; 2492: 38; 2493: 36; 2498: 188, 190
V 2 3: 192
XI 2 142: 197; 145: 187; 162: 196; 287 A: 194–5
XII 5 2: 191
XII suppl. 349: 139; 347 II: 146

IJG
(Dareste, R./Haussoullier, B./Reinach, T. (1891–1904), *Recueil des inscriptions juridiques grecques*, Paris); 12: 188; 13: 188; 19a: 148

IP
(Hiller v. Gaertringen (1968), ed., *Die Inschriften von Priene*, Berlin, first edn 1906) 3.12–14

Isaeus
6.33: 101; 8.15–16: 178; 11.8–9, 41: 127; 11.49–50: 128; 12.2: 128

Isocrates
Archidamus 96: 169
Panathenaicus 178: 152
Panegyricus 31: 172

LSCG
(Sokolowski, F. (1961), *Lois sacrés des cités grecques*, Paris) no.67: 192; no.105: 191; no.116: 192

Lysias
7: 118, 155; 7.3, 32, 41: 204; 7.4–5, 9–10: 155; 19.29: 78; 22: 21, 106

Meiggs/Lewis 1969
no.58: 140; no.67: 142, 143; no.69: 171; no.73: 171; no.84: 174; no.89: 173

Menander
   Dyskolos 377: 81

Pausanias 130
   1.32.7: 15
Photius
   s.v. *kallikyrioi*: 152
Plato
   Alcibiades 1.123c: 78
   Critias 111c: 12
   Laws 118; 547c: 152; 738b–c: 185;
      738c–d: 182; 738d–e: 184; 809d:
      164; 842e–846c: 147; 847e: 139,
      140; 848a: 141; 625d: 132; 739–40:
      126; 945e: 163; 956d–e: 137
   Republic 118; 416d: 123
Pliny
   Naturalis Historia 2.89: 141
Plutarch 130
   Cimon 13: 106
   De Genio Socratis 580E: 103
   Lycurgus 8.3–4: 132; 8.7: 151; 15.7:
      132; 16.1: 132, 133; 24.2: 151
   Moralia 239d–e: 15
   Nicias 3.6: 183
   Pericles 12: 105
   Solon 24: 146; 31.5: 145
   Quomodo adulator ab amico
      internoscatur 143
Polybios 130
   6.45.3: 132; 6.54.3: 131
Porphyrius
   Abstinentia 2.24: 169

SEG
   1.366; 24.151: 129
SIG^3
   (Dittenberger, G. (1915–24), *Sylloge*
      Inscriptionum Graecarum (3rd
      edn), Leipzig) 141: 124; 153: 195;
      175: 193; 178: 193; 407: 192; 636:
      192; 826 G: 193; 963: 24, 191;
      965: 24; 976: 172; 986: 192; 1264:
      163
Sophocles
   Oedipus Coloneus 695 ff.: 203
   Oedipus Rex 1121 ff.: 99
Stephanus of Byzantium
   s.v. *Neai* 141

Tertullian
   Apologeticus 13.6: 135

Theognis
   183 ff.: 101; 863 ff.: 95; 1112: 101
Theophrastus 38, 199
   Characters 9.1–3: 179
   De Causis Plantarum 7; 1.6.10: 35;
      1.20.3: 40; 2.9.5: 41; 3: 38; 3.2.6:
      29; 3.4: 29; 3.5.5: 35; 3.7.7: 39;
      3.8.1: 39; 3.9.1: 40; 3.9.5: 32; 3.10:
      32; 3.10.8: 29; 3.11 ff.: 28, 29;
      3.13: 32; 3.20.1: 24, 49; 3.20.6: 25;
      3.20.8: 49; 4.8.1: 42; 4.11.4: 24;
      4.13.3: 25; 5.1.3: 28; 5.4.2: 40
   fr.97 128
   Historia Plantarum 7, 28; 2: 38;
      2.1: 29, 35; 2.5: 29; 2.5.4: 35;
      2.7.2 ff.: 38; 2.24: 29; 3.5.4: 32;
      4.13.2: 38; 4.13.5: 38; 5.3: 29;
      5.9.8: 35; 7: 43; 8.1.1: 21, 42;
      8.2.5: 42; 8.2.6: 42; 8.2.7: 25;
      8.4.4: 24; 8.5.1: 42; 8.6.1: 24;
      8.7.3: 42; 8.9.1: 42, 110
Thucydides
   1.10.2: 131; 1.13.6: 183; 1.80.4:
      142; 2.14: 69, 102; 3.50.2: 183;
      3.104: 187; 3.104.2: 183; 5.42:
      100; 7.28.4: 144; 8.38: 72; 8.40:
      72
Tod 1948
   no.107: 106; no.162: 141; no.167:
      144; no.196: 140; no.198: 104;
      no.204: 120
Tyrtaeus
   fr.6 151

Varro
   De Re Rustica 94–5; 1.28: 168

Walbank 1983
   Stele I 183

Xenophon 38, 130
   Anabasis 118; 1.5.10: 25; 5.3.4–13:
      173–4
   Cyropaedia 118
   Hellenica 118; 3.3.6: 153; 4.4.6: 120;
      6.4.29: 175
   Hieron 4.2: 173
   Lacedaimonion Politeia 5.3: 138; 15.3:
      131
   Poroi 142
   Oeconomicus 7, 118; 1: 127; 1.9: 84;

3.10.1–3: 153; 5.10: 173; 5.16.1–2: 153; 5.18–20: 163; 6.8–10: 149; 9.3: 67; 16.9: 21; 16.10: 24; 16.15: 49; 17.2: 163; 17.6: 21; 17.12 ff.: 25; 18.1 ff.: 25, 53; 18.3: 84; 19.1 ff.: 27–8; 19.13: 35; 20.10 ff.: 111; 20.22–6: 128
Symposium 4.49: 169

GENERAL INDEX

Achilles, shield of 22, 25, 26, 47, 52, 56, 98
achyron 197
Acropolis 38, 68, 203
adiairetos 124
adoption 127–8; of girls 127
Aegina 61
Aetolia 171
agricultural year 104, 161–2
agronomoi 137, 147
agros 80
Agyrrhios 140–1
Alcibiades 78
Alea Athene 191–2
alfalfa 17, 111
Alkinoos 26, 41
almond 41
aloe 53
Amorgos 24, 183, 191
Amphiaraos 186
amphorae 201; Panathenaïc 204–5; SOS 201
anadasmos tes ges 128
anadendras 29, 36
Androtion 39–40
animals: castration of 87–9; domestic 85, 102; of the gods 191–8; living in herds 85, 96; mating of 85, 96; sacrificial 85, 107, 174–80, 193, 198; trade in 107
anonyma 42
Anthesteria 57, 162, 166
antidosis 141
aparche 136, 169–73
Aphrodite 166, 183
Apollo 164, 173; animals of 192–8; festivals of 162, 165–6, 170
apple 10

Arcadia 191
Archidamian War 102
Arcturus 26, 32, 162
ard 46, 49–50, 56, 110; modern 47–8
Argolis 5, 100
Aristotle, ideal state of 122, 150
arpe 25, 52
Arrhephoria 162, 166
Artemis 173–4, 184–5; festival of 162, 166, 167
Artemis Orthia 53
Asia Minor 11
Athene 118, 173, 178, 188; festivals of 162, 166–7
Athenians 100, 102
Athens 126–9, 153–5; *passim*
Attica 203–5; *passim*
aula 98, 100
autogyon aratron 46

barley 21, 25, 32, 140, 174; two-rowed 22, six-rowed 22
beans 32, 42
bees 96, 103, 148
beets 43
Berezan 201
bird-cherry 42
birth control 126
blastologia 32
Boeotia 5, 13, 14, 47–8, 98, 100, 161, 164–5
border-area 100, 122, 125, 129, 185; conflict 100
Bosporus 144
bothros 28
bothynos 28
boustasis 196
Brea 184

building *see* construction year; transport

cabbage 43
cadastre 142
calendar 160–8; agrarian in Hesiod 7, 22, 26, 161, 168; agrarian in Varro 168; Attic of festivals 160, 162, 165–7; of demes 175–80; frieze 167–8; sacred 163–5
cavalry 86
celery 43
chamakes 32
charakes 32
Charoneia, at Delos 196
cheese 91, 101, 103, 138
chestnut 41
chick pea 42
Chios 11, 31, 60–1, 71–2, 81, 192
Chiron 177
chorion 36, 80, 101, 139
chortos 25
cicer arietinum 42
citrus fruits 6
cleruchies 140
climate 10–11, 13
cockfighting 95
colonies 122, 123–6
colonization 9, 110
common messes 122, 132, 137–40, 150
construction year 104, 161
Corinth 5, 100, 114, 120; Corinthian War 100
cow *see* oxen
cowshed 76, 196–7
Crete 123, 132, 137–9, 144, 150–3, 163, 188
Crimea 69–70, 76–8, 144
crops, perennial 26
Cyclades 11, 13
Cyclopes, island of 27
Cyrus 173
cytisus 103

day-labourer 154
decree *see* laws
deforestation 12–14
dekate 136, 139–40, 170, 173–4
Delos 66, 107, 171, 180, 181, 183, 186–90, 193–8; press-room in 68–9
Delphi 165, 171, 175, 180; accounts 192; Oracle of 185–8

demes of Attica 69, 129, 141, 165, 187–90
Demeter 55, 163, 170, 172, 185; festivals of 24, 162, 166
Dicaearchus 138
digging 27, 32
dikella 49
dimenoi 24
Dionysia, rural 165
Dionysos 26, 166; festivals of 162
dog 85, 91
donkey-mill 106
donkeys 85–9; as carriers 55; as draught-animals 17; *see also* transport
douloi 150, 153
dowry: rules of 17, 127; security for 141, 183;
draught-animals 17, 25, 46, 84, 102, 104, 192; *see also* donkeys; mules; oxen; transport
drepane 52
drepanon 52
dryinon 80
dung *see* manure

echetle 46–7
Egypt 150
eikoste 136
eiresione 170, 172
eisphora 104, 135–7, 141–3
Eleusinian Mysteries 171, 175
Eleusis 24, 104, 105
Elis 98
elyma 46–7
elymos 42
enerosia 196
enneaboios 97
epaulos 100
epikarpia 137
epikleros see heiress
epinomia 100
Epiros 98
epistates 153
Epitadeos 133
Erchia, calendar of 175–80
erebinthos 42
Erechtheion 203
ergatai 153
erosion 11, 14
eschatia 72, 79, 81
Euboea 5, 56, 78, 98, 102
Eudemos of Plataea 104–5

GENERAL INDEX

eukampes 52
Euxenippos 186
exairetos 124, 182
export of foodstuffs 117–18

fallow 22, 24, 43, 49, 108–9, 112, 161; of the horse mare 87
farmers: dependent 151–2, 154, 188; place of habitation 67–82, 101–2, 150, 153; subsistence 113
first crop 169–74, 180
first-fruits offering *see* first crop
foreigner 143
farmsteads: 'agro-pastoral' 102–3, 110; evidence for location 67–82; size of 71
fertilizer 17
fertilizing 39–40
fig 41, 103, 138, 162, 197
fodder 25, 84, 86, 87, 101, 103, 105–6, 143; for horses 52; plants 17, 108–10, 112
fruit, dried 174
fruit trees 26, 36, 41–2, 193, 197
fuel 103, 106

garlic 43
ge: *demosia* 121; *hiera* 121; *idia* 121; *koine* 121; *psile* 80
Gela 71
Genesia 162, 167
georgos 153, 155
gepedon 80
goats 17, 91–3, 101–4, 192; for sacrifice 93, 175–6
goose 95–6
Gortyn 119, 123, 139–40, 148, 150–1
grain 21–6; chaff 25; harvest 25, 162; quern 106; straw 25; supply 139–40; threshing 25, 162; transport 106; winnowing 25; yield 26
gye 46–7
gymnasia 150
gyros 29

haimasia 81
Halieis 68, 125
hammer 56
harness 85
harvest *see* grain
harvest festival 166
hazel 41

hecatomb 174, 188
heiress 127, 151
hekatomboios 97
Helos Plain 6
helot 5, 133, 151–3, 199
hemeros 85
hen 94–5
Hera: festival of 162; land of 172
Heraion 173
Heraklidai, return of 20, 129–30, 133
herdsman *see* shepherd
Hermogenes 169–70
hetaira 173
hiera syggraphe 194, 196–8
Hieron II 175
hieropoioi 194–5, 197
hierothytes 192
hippeis 85
hippobotai 85
Hippodamos, ideal state of 122
Hippodromos 197
hippothelai 87
hired worker 100
histoboe 46–7, 56
hoe 49–52
hoeing 24, 25, 39
Homer, similes in 7, 97
honey 96
hordeum 21
horoi 154–5, 183
horses 52, 85–9, 101, 111, 177, 192–3
hynis 46–7
hyperon 55
hypokonisis 32
hypolenion 57
hypomeiones 134, 153
hypopremna 35
hypozygia 102

Imbros 140
import of foodstuffs: 117–18; of grain 106; of wood 12
improvement, of soil 42–3
inheritance, rules of 18, 118, 126, 150–1
intercalation 171
intercultivation 32
Ios 191
iron spits 53, 138, 173
irrigation 5, 17, 39–40, 52, 112, 128
Issa 124

Jason 175

GENERAL INDEX

Kallikles 148
Kallikrates 183
kallikyrioi 152
karpodaistai 139
kartaipos 148
Karthaia 181
Kavalla 173
kegchros 21, 42
Keos 6, 71, 80, 103
kepos 80
Kerameikos 178
Kerkyra Melaina 124–5
Kirrha 186, 192–3
Kleisthenes 129
klema 27
kleros 69; *see also* land
klisiai 98
knives 44, 52–3
Knossos 148
kophinoi 56
kopros 98; *see also* manure
kotinos 35
krithe 21
Kronia 162, 166
Kroton 125
kyamos 42
Kyrene 140

Laertes 27, 41, 81, 97
land: confiscation 128, 129; distribution 124; fragmentation 128; marginal 103, 110; private, 120–34; public 102, 121–3, 154; redistribution 128; right to own 104, 149; sale of 123, 124–5, 128, 129, 141; size of lots 69, 78–80, 126, 132; status of 121–3; surplus 122–3, 129; terminology 80, 121
land division, in Aristotle 122
land of the gods 36, 119, 121–3, 154–5, 172, 181–90; administration 186–8; leasing 188–9; tenants 189–90, 193–8
laws and decrees on: *aparche* 170; *argia* 145; bees 148; citizenship 127; export 145; farmers in Plato 147; grain tax 140; import 146; inheritance 126, 133, 150–1 (*see also* inheritance); leasing 119; sacred olives 118, 204; trade 117; water-course 118, 147, 148
leaseholders 154
leasing 24, 36, 119, 123, 129, 132, 141, 154–5; accounts 119, 196–7; contracts 79; lists 119; *see also* land of the gods

Lemnos 129, 140–1, 172
Lenaion 164
lenos 56
lens esculenta 42
lentil 25, 42
Lesbos 11, 107, 183
lesche 67
liknon 55
Linear-B tablets 19
lions 97
liturgy 141, 179, 200–2
Lokris 100
Long Walls 105
lucerne 111; *see also* alfalfa
lupin 25
Lykourgos 132–3
Lysimachos, son of Aristeides 78
Lyttos 139

Macedonia 12–13, 42, 110; Philip of 186
Magnesia 146, 148
Maimakterion 162, 168
maize 6
makele 49, 52
makella 49, 52
manure 5, 24, 25, 108–9, 111
maquis 14, 36, 111
Marathon 176; Plain of 14–18, 98
maritime suits 146
mechanization 5
medimnos 138
Mediterranean triad 20
Megara 98, 103, 185
Melos 6, 71
Mende 142–3
merchant 21
mere, ta 124
mesaulos 98
Messenia 5, 151
Metapontum 23, 69–70, 71, 76
Methana 16, 18, 30, 34, 53, 61, 81, 103–4, 128
metic 153–4, 189
miliarium 60
milk 89–91, 97, 103, 104, 111, 195–6
mill 106, 192
millet 21, 42; Italian 42
mining 201
minors 141, 154–5
mixed farming 15–16, 102, 108, 196–8
mola olearia 60–1
moriai 203

mortar 55, 192
mortarium 60
moscheuein 29
mothakes 134
mules 85–9, 101; as carriers 55; as draught-animals 7, 22, 47, 50–1; *see also* transport
Mykonos 47, 193, 197
myrtle 39

Nausithoos 9, 70
Nea 129, 141, 188
neao 49
neios 22
neodamodeis 134, 153
Niketeria 162, 167
Nikias 183, 197
Nikosthenes, painter 49
nomos 103
nursery 4, 27, 36

oath, of the recruits 120
obeloi see iron spits
oikeis 150–1
oiketai 143, 153
oikia 80
oikistion 53
oikopedon 80
oikos 127–8
Olbia 201
olea: *chrysophylla* 33; *europaea* 33; *sylvestris* 33
olive 26, 33–40, 138, 140, 166; as fodder 103; as fuel 103; not in Hesiod 7; harvest of 38, 40, 57–60; implements for making oil 57–66; oil 20, 33; planting and propagation of 35–8; pruning 38–9; sacred 38, 118, 203–5; wild 33, 34, 36, 37, 38; *see also kotinos*; *olea*
olmos 55
Olympia 38,
Olympos, Mount 11
Olynthos 53, 59, 60, 61, 62
onions 43
orbes 60
orchard 41
orgas 80
Orgas, the Holy 185, 187
orgeones 190
Orion 162
Oropos 129, 141, 172, 178–9, 186

oryttein 49
Oschophoria 162, 166
oxen 89–91, 102, 192; branded 194; as draught-animals 47, 50–1, 53, 89, 102; in Hesiod 7, 46; hides from 179; for sacrifice 91, 171–2, 174–6; *see also* transport

paddock 197
pala 49
Panaitios 102–3
Panaktos 100
Panathenaia 162, 166, 168, 171, 174, 178–9, 188, 204
panicum miliaceum 42
panoply 171–2
Parthenon 104
Parthenos 173
pastas 150
pea 42
pear 41
peasant 113–14
Peisistratos 143, 145
pelanos 172
penestai 152
pentekoste 136, 140
Perachora 53
Perinthos 166
perioikoi 131, 133, 142, 150, 152–3
Persephone 172
Persia 111
phakos 42
Phocaea 100
Phormisios 79
phormoi 56
phyta, ta 27
phyteia, ta 28
phyteuein 28
phyteuterion 29, 36
phytorion 38
pig 85, 93–4, 103–4; for sacrifice 174–6, 195
pigeon 94
Pindos Range 11
Piraeus 24, 140, 188–90
pisos 42
pisum sativum 42
pityinon 80
Pleiades 22, 24, 32, 162
plough 46–9, 56
ploughing 22, 162, 168; in of beans 42; thrice-ploughed field 22, 24, 49, 162

plum 42
poasmos 25
pollen 12–14
Polyeuktos 186
Polykrates 183
polymelos 84
pomegranate 41
Pompeii 62
Poseidon, festivals of 162, 195
poultry 94–6
Praisos 61, 63, 66, 68
price, of: cheese 103; herds and shepherd 101; hides 179; land 183
Priene 125
probata 102, 103
promoscheuein 29
property: invisible 142; visible 142
ptyon 55
pulse 17, 42–3, 103, 108, 110, 112; threshing of 55
Pyanopsia 170
Pyanopsion 162, 168, 172
pyros 21
Pythia 175

quince 41

rabdizein 40
rainfall 10–11, 25, 99, 147–8, 163
Rheneia 183, 187; tombs of 53
Rhodes 11, 81
risk, minimizing of 18
ritsosis 29
Romans 4, 112, 135, 143
rotation of crops 17, 24, 25

sacred: animals *see* animals of the gods; land *see* land of the gods
sacrifice 160; of animals 174; distribution of meat 177–9; financing of 179–80
Salamis 129, 172, 183
Samos 11, 70, 166, 172, 173
Sarakatsani 99
Scheria 9, 26
screw: Archimedean 112; press 4, 63
scrobes 29
secularization 187
sekoi 98
Selymbria 146
seminarium 27, 38
setaria Italica 42

settlement 6
sheep 17, 91–3, 102–4, 192; associated with agriculture 84; lambing 101; for sacrifice 93, 175–6; shearing 101
shepherd 91, 92, 97, 98, 99–102
Sicily 71
sickle 25, 44, 52–4, 56
Siphnos 70
Sirius 32, 162
sitos 21
skalis 52
skalsis 25, 52
skaptein 49
skene 100
Skillus 173, 180, 181, 183–5
Skyros 140
slaves 133, 166, 196; agricultural 110, 153–4, 199; artisan 101; shepherd 100
sminye 52
Smyrna 143
Solon 109, 128, 145, 154; *tele* 79, 136
solstice 162–3, 165; summer 163; winter 26
Sounion 53, 68
sowing 22, 24, 47–9, 50–1; implements 52
spade 49–52
Sparta 5, 38, 124, 129–34, 137–9, 142–3, 149, 150–3, 188
sphyra 56
spices 43
stadion 104
stars 162–3
stathmos 98, 100
stock-breeeding 17
storage 55, 68; silos for 55
sundial 163
Sybaris 125
synanthropeuomena 85, 93
synoikia 80
synoikismos 69, 166
Syracuse 175
syrphetos 25
syssitia see common messes

Tartessos 173
tax 200; collector 140; evasion 142; indirect 143–4; on land 135–6, 141–3; in Plato 136–7, 141, 143–4; poll 138; on produce 137–41, 143–4, 196; property 137, 141–3; purchase 143; terminology 136–7

GENERAL INDEX

Tegea 191
Teithras 129
telos 136, 173
temenos 182
tenant 24
terrace 4, 72, 81–2
Thales from Miletos 65–6
Thargelia 162, 166
Thasos 71, 136, 139, 146–7
Thebes 63
therina 42
Thespiai 181
Thessaly 42, 98, 110, 152, 175
thremmata 191–2
threshing *see* grain
threshing floor 25, 53–5
threshing sledge 53, 56
Thourioi 125
thysia 174, 177
tithe 135–6, 139, 170, 173–4
tobacco 6
towers 68–9; Princess 53; Cliff 55
trade, scale of 200–2
transhumance 17, 99–101, 191–2
transport 55, 104–7
trapetum 60–2, 66, 68
tree-medick 103
trees, propagation of 27–9
tribolos 56
trierarchy, Athenian 200
trimenoi 24
tripolos 22
Triptolemos 23, 24
triticum 21; *cereale* 21; *dicoccum* 21; *vulgare* 21
turnips 43

Vari: country house of 73–4; country house near 74–5
vegetables 43
vetch 25
vicia faba 42
villae rusticae 4, 80
villaticae pastiones 94
vine 26–33, 41, 193, 196–8; cuttings of 30; planting of 27–9; pruning 26, 32, 39, 162; staking 32; vineyard 26–7, 30; *see also* wine
Vlaches 99

waggon 55
walnut 41
war, spoils of 170, 173
watering *see* irrigation
weeding 25
wheat 21, 25, 140, 174; emmer 21
wine 138, 146, 174; implements for making 56–9, 63; pressing of 8, 26, 56–9, 68; vintage 8, 26, 32, 56, 162; *see also* vine
winnowing: basket 55; shovel 55; *see also* grain
women 127, 131, 150–1, 154, 173, 178, 183
wool 91, 97, 193, 195

Xenophon 173–4

year: military 167; new 165–6; *see also* agricultural year; construction year

zeia 21
zeugos 105
Zeus 163–4, 191; Ctesios 178; Epopsios 163; festivals of 162
zygon 46